Nanomaterials for Applications in Energy and Biology 2016

Topical Conference at the 2016 AIChE Annual Meeting

San Francisco, California, USA
13-18 November 2016

ISBN: 978-1-5108-3421-7

Printed from e-media with permission by:

Curran Associates, Inc.
57 Morehouse Lane
Red Hook, NY 12571

Some format issues inherent in the e-media version may also appear in this print version.

Printed by Curran Associates, Inc. (2017)

AIChE
120 Wall Street, FL 23
New York, NY 10005-4020

Phone: (800) 242-4363
Fax: (203) 775-5177

www.aiche.org

Additional copies of this publication are available from:

Curran Associates, Inc.
57 Morehouse Lane
Red Hook, NY 12571 USA
Phone: 845-758-0400
Fax: 845-758-2633
Email: curran@proceedings.com
Web: www.proceedings.com

TABLE OF CONTENTS

460852 Engineering Macroscale Thermoelectric Transport By Chemical Modulation of Nanoscale Interfaces

Abstract:

Optimizing the performance of thermoelectric materials has historically proven difficult as their essential transport parameters are cross-coupled in bulk transport equations. However, new classes of hybrid inorganic/organic materials are emerging which present exciting opportunities as their transport behavior can strongly deviate from classical models. In this talk, I will present recent achievements in demonstrating new, interface-driven mechanisms for enhancing and modulating thermoelectric transport in nanoscale hybrid systems. Highlighted research will include p- to n-type conversion of transport type in Tellurium nanowires, subphase alloy scattering to boost power factor in Tellurium nanowires, phase transition anomalies in Cu_2Se nanocrystal arrays, and unexpectedly low thermal conductivity in hybrid perovskite nanowires.

472306 Doped Semiconducting Polymers As Solution-Processable Thermoelectric Materials

Abstract:

Due to inherently low thermal conductivity and tunable electronic properties, semiconducting polymers are very promising solution-processable-materials for thermoelectric applications. Polymeric semiconductors are insulators in its pristine state, but are rendered conductive through chemical doping. Here, we studied the thermoelectric properties of poly(2,5-bis(3-tetradecylthiophen-2-yl)thieno[3,2-*b*]thiophene (PBTTT-C_{14}) doped either from solution or from the vapor phase. Overall, vapor doped films yield large Seebeck coefficients at high electrical conductivity (Power Factors ~ 100 Î¼$Wm^{-1}K^{-2}$). It important to note that the introduction of the chemical dopants can perturb the molecular packing and the morphology relative to the neat film. By leveraging various synchrotron X-ray scattering techniques that probe both the local and long range order, we discovered key structural features that can be correlated to the enhancement in the Seebeck coefficient and electrical conductivity for vapor doped films. Specifically, the introduction of the chemical dopant must limit the structural perturbation at small length scales while maintaining long-range correlation of polymer chains essential to efficient macroscopic charge transport. Lastly, to further elucidate the charge transport mechanism driving the thermoelectric properties, we report on the temperature-dependent measurements of both the Seebeck coefficient and electrical conductivity.

468413 Thermal and Thermoelectric Transport Coefficients in Graphene

Abstract:

Thermal and thermoelectric transport coefficients

in graphene

Enrique Muñoz, Physics Institute, Pontificia Universidad Católica de Chile, Vicuña Mackenna 4860, Santiago, Chile.

Abstract

Thermal, electronic and thermoelectric transport coefficients in graphene were studied. The effect of electron-phonon scattering processes, as well as the presence of C^{13} isotopes over the thermoelectric properties of extrinsic graphene were considered. Electron-phonon interaction is described within the deformation potential approximation. Electrical and thermal resistivity, as well as the thermopower, were calculated within the Bloch theory approximations. Analytical expressions for the different transport coefficients were obtained from a variational solution of the Boltzmann transport equation. The phonon-limited electrical resistivity $\rho e{-}ph$ shows a linear in temperature dependence at high temperatures, and follows a $\rho e{-}ph$ T^4 at low temperatures, in agreement with experiments. The phonon-limited thermal resistivity at low temperatures exhibits a T dependence and achieves a nearly constant value at high temperatures. The predicted Seebeck coefficient at very low temperature is $Q(T)$ $n^2 kB\,T/(3eEF)$, which shows a $n^{-1/2}$ dependence with the carrier density n, in agreement with experiments. On the other hand, the presence of a random distribution of C^{13} isotopes at fixed concentrations was studied as an strategy to reduce the lattice thermal conductivity, with the corresponding enhancement of the figure of merit for

thermoelectric applications.[3]

[1] E. Muñoz, Journal of Physics: Condensed Matter 24 (2012), 195302.

[2] E. Muñoz, J. Lu and B. I. Yakobson, Nano Letters 10 (2010), 1652.

[3] E. Muñoz, submitted (2016).

463547 Carrier Scattering at Alloy Nanointerfaces Enhances Power Factor in PEDOT:PSS Hybrid Thermoelectrics

Abstract:

This work demonstrates the first method for controlled growth of heterostructures within hybrid organic/inorganic nanocomposite thermoelectrics. Using a facile, aqueous technique, semimetal-alloy nanointerfaces are patterned within a hybrid thermoelectric system consisting of tellurium (Te) nanowires and the conducting polymer poly(3,4-ethylenedioxythiophene): poly(styrene-sulfonate) (PEDOT:PSS). Specifically, this method is used to grow nanoscale islands of $Cu_{1.75}Te$ alloy subphases within hybrid PEDOT:PSS-Te nanowires. This technique is shown to provide tunability of thermoelectric and electronic properties, providing up to 22% enhancement of the systemâ€™s power factor in the low-doping regime, consistent with preferential scattering of low energy carriers. This work provides an exciting platform for rational design of multiphase nanocomposites and highlights the potential for engineering of carrier filtering within hybrid thermoelectrics via introduction of interfaces with controlled structural and energetic properties.

465380 Combining Density Functional Theory Calculations, Supercomputing, and Data-Driven Methods to Design New Thermoelectric Materials

Abstract:

Density functional theory (DFT) simulations solve for the electronic structure of materials starting from the Schrödinger equation. Many case studies have now demonstrated that researchers can often use DFT to design new compounds in the computer (e.g., for batteries, catalysts, and hydrogen storage) before synthesis and characterization in the lab. In this talk, I will focus on how DFT calculations can be executed on large supercomputing resources in order to generate very large data sets on new materials for functional applications. First, I will briefly describe the Materials Project, an effort at LBNL that has virtually characterized over 60,000 materials using DFT and has shared the results with over 17,000 registered users. Next, I will talk about how such data can help discover new materials, describing how preliminary computational screening led to the identification and confirmation of a new family of bulk AMX2 thermoelectric compounds with measured zT reaching 0.8. I will outline future plans for how such data-driven methods can be used to better understand the factors that control thermoelectric behavior, e.g., for the rational design of electronic band structures, in ways that are different from conventional approaches.

450621 Phase Diagram, Microstructure and Thermoelectric Properties

Abstract:

Thermoelectric materials and devices have attracted intensive research and development interests due to their ability to convert waste heat directly into electricity. Energy usage efficiency could then be improved which is a very critical concern nowadays. Good thermoelectric properties require low thermal conductance but good electrical conductance. Phonon scattering caused by nanoprecipitates has been reported to be an efficient mechanism to enhance the thermal resistance but causing no significant hindrance on electrical conductance. Microstructures of thermoelectric materials are essential for their thermoelectric properties. Phase diagrams provide information of phase equilibria, and are fundamentally important for processing routes selection and understanding of materials microstructures.

Ternary thermoelectric materials systems, Bi-Te-In, Ag-Pb-Te and Co-Sb-Ga, are used as examples to illustrate the relationships between phase diagrams, microstructures and thermoelectric properties. Two kinds of ternary alloys were prepared. Ternary alloys of one kind were equilibrated at pre-determined temperatures, and the equilibrium phases were determined. Alloys of the other kind were quenched, and their primary solidification alloys were determined. Isothermal sections of ternary phase equilibria were proposed based on the results of determined ternary equilibrium phases together with the information obtained from their constituent binary phase diagrams, while the liquidus projections were determined based on results of the primary solidification phases.

According to the determined liquidus projections, ternary alloys with compositions at the ternary eutectics in the Bi-Te-In and Ag-Sb-Te systems, L (Bi- 13.0at%In-77.0at%Te)=Bi_2Te_3+Te+In_2Te_5, and L (Ag-40.at%Sb-36.at%Te)=$AgSbTe_2$ +Ag_2Te+δ-Sb_2Te, were prepared. Their microstructures were determined, and their thermoelectric properties were measured. Nanoprecipitates of Ag_2Te phase were observed. This ternary eutectic Ag-Pb-Te alloy has very low thermal conductivity (0.3w/mK at 415K), presumably due to the nano-sized lamellar structure and the nanoprecipitates enhancing phonon scattering. Very fine microstructures with nano- whisker-type precipitates were observed in the as-solidified ternary Bi-Te-In eutectic. These nanoprecipitates contribute to the alloys' thermoelectric properties.

Ternary alloys in the Bi-Te-In and Co-Sb-Ga systems with compositions close to the ternary solubility limits of Bi_2Te_3 and $CoSb_3$ phases were also prepared. The thermoelectric properties of the Ga-doped $CoSb_3$ alloys were improved with the concentration of Ga doping and the figure of merit increased to 0.7 with a 0.1at% Ga doping. Similar results were observed in the In-doped Bi_2Te_3 alloys. A small amount alloying would enhance the phonon scattering and reduce lattice thermal conductivity. However, secondary phases would form if the doping concentrations are higher than the solubility limits. Since the secondary phase might not have good thermoelectric properties, its existence would deteriorate the thermoelectric properties of the alloys. The phase diagrams are thus fundamentally important.

459105 Shaping the Spectrum of Thermal Radiation Using Nanostructured Materials for Efficient Thermophotovoltaic Power Generation

Abstract:

By manipulating the structure of materials at the nanoscale, we can tune the spectrum of thermal radiation. In doing so, we can selectively enhance or suppress certain modes linked to either desirable or undesirable properties. Such spectrally-selective transport holds promise for more efficient solar and thermal power generation.

Here, we enhance spectrally-selective radiative transport using nanostructured materials to improve the efficiency of thermophotovoltaics (TPVs). TPVs are solid-state heat engines that convert energy from a high temperature radiant source—the emitter—into excited charge carriers in a semiconductor diode—the cell. TPVs are a promising technology for applications in concentrated solar power and distributed combined heat and power because they are solid-state, scalable and compatible with a variety of high-temperature heat sources. To achieve high efficiencies, it is well known that spectral control over thermal radiation is critical. Nevertheless, several questions remain: 1) *What spectral features are most important in TPVs when accounting for parasitic losses and non-idealities?, and 2) How can nanostructured materials be designed to meet the stringent spectral and thermal requirements in TPVs?*

To address these questions, we developed radiative transport and energy conversion models to identify the most important spectral features in realistic TPVs. Based on these models, we show that suppressing radiative exchange between the emitter and the cell at energies below the bandgap of the cell (i.e., at long wavelengths) is critical for high efficiency. This is explained by the fact that, at the optimal operating temperature, the emissive power below bandgap is much higher than above the bandgap. Operation at higher temperatures is undesirable because of the increased high-temperature radiative losses and excited charge carrier thermalization in the cell.

Based on these insights, we developed thin-film and nanostructured materials for TPV applications where spectrally-selective transport is based on the interaction of thermal radiation with wavelength-scale structures. The structures are chosen to minimize radiative transport at energies below the bandgap and then optimized to achieve high efficiencies in the presence of realistic losses. By suppressing radiative exchange at low energies, we demonstrate a solar TPV device that is an order of magnitude more efficient than previous solar TPVs and has the potential to exceed the limit of single-junction solar cells. Further, we show how a scalable and thin-film cell design promotes spectral selectivity using a back-side surface reflector to return sub-bandgap radiation back to the hot emitter. Spectroscopy measurements confirm that the reflectance below the bandgap for these thin-film cells is significantly higher than for cells on bulk substrates. Based on these spectral properties we show that the heat engine efficiency can be significantly higher than the state-of-the-art, exceeding 30%.

This work provides a detailed understanding of the relative importance of spectral features in TPVs and offers strategies for how to best engineer thin-film and nanostructured surfaces to shape the spectrum of thermal radiation for TPVs. With integration of thermal storage, TPVs can become an important part of distributed energy systems and may lead to increased adoption of intermittent renewables.

469139 Application of Metamaterials and Rectenna for Capture of Blackbody Radiation

Abstract:

Metamaterials, a new synthetic set of materials with properties that do not exist on their own in nature, typically consist of various composite materials such as metals and dielectrics. Their repeating geometric structure and material set provide the metamaterial with a negative index of refraction. The negative index of refraction induces a primary function of metamaterials: their ability to manipulate electromagnetic waves. In this study, metamaterials are designed and fabricated to manipulate blackbody radiation into a well-organized, functional source of electromagnetic waves that are rectified to a DC voltage.

The metamaterial-rectenna unit is designed to convert heat into electricity using the optical properties of the blackbody radiation emitted from heat source. The blackbody radiation is optically manipulated by the metamaterial. In general, as the temperature of the object increases, the wavelength and intensity of the blackbody radiation increase. The metametarial-rectenna units fabricated in this study are designed to capture electromagnetic radiation with a frequency of 1 THz.

The constructed device uses a rectenna as the rectifying device to convert the alternating current (AC) induced from the electromagnetic waves emitted as blackbody radiation to a direct current (DC). The rectenna, containing a metal insulator metal (MIM) tunneling diode, utilizes the ordered electromagnetic radiation from the metamaterial. The metamaterial consists of copper and a photosensitive polymer in an alternating pattern that focuses the electromagnetic radiation over onto the rectenna. The metal of choice, copper, is selected due to its high thermal and electrical conductivity. SU-8 is selected as the photosensitive polymer due to its well-characterized optical properties as well as use in photolithography. The metamaterial is fabricated using various techniques of photolithography, including wet and dry etching of metals and oxides, thin-layer deposition of metals using evaporation, and electrodeposition.

In order to test the functionality of the metamaterial, a heat source applies heat at a set temperature to the metamaterial. Using test pads that have been built into the structure and connected to the rectenna, the diodes are characterized using four-point probe IV measurements. The thermal measurements utilize two of these probes to detect any voltage that is generated by the rectenna-metamaterial device when exposed to a controlled heat source. In this study, a proof of concept design of the metamaterial-rectenna units is achieved and a plan for future development is suggested.

469647 Testing Materials and Devices for Electromagnetic Energy Capture and Conversion to Electricity

Abstract:

The subject of this paper is terahertz (THz) recteenna. Rectenna are devices based on two fundamental components. There is an antenna that captures the photonic energy and a rectifier that transforms alternating current (AC) into a directional current (DC) for use in power systems. Research on our team is presently focused on developing efficient rectenna devices capable of capturing electromagnetic (EM) radiation is the form of infrared (IR) emitted waste heat. Materials-based solutions used in this regard include tailoring antenna geometry and composition and coupling metamaterials into the receiver. The antenna are coupled through a feedpoint to a metal-insulator-metal (MIM) type diode. MIM diodes possess the characteristic of being well-suited for high-speed aplications.

This bulk of this paper will center on testing strategies which are not trivial in the 30 - 300 THz region. Both free-space testing and thermal testing results will be presented. In order to truly quantify overall efficiency numbers, our reserach team uses computational finite element modeling (FEM). Thie paper will present results from FEM thermal transport models, optoelectronic models, and nanofabrication models. In summary, results will be provided about rectenna device performance at 1, 5, and 30 THz.

459954 Incorporation of Photo-Responsive Membrane Protein Species into Nanostructured Silica for Light-Driven Ion Transport

Abstract:

Membrane proteins are versatile biomolecules with diverse functionalities that impart sensing, signaling, transport, or catalytic properties that support the viabilities of biological cells. Such functionalities are highly selective to particular ions or molecules and often occur at high rates, which would be attractive for technological applications, such as chemical or biological sensing, separations, bioanalytics, and energy conversion. To effectively exploit membrane proteins for technological purposes often requires their incorporation into synthetic host materials that enable the proteins to function stably and be integrated into macroscopic devices. One interesting example is the membrane protein proteorhodopsin, which functions as a light-driven H^+-ion-pump that might be harnessed for photochemical energy conversion. Synthetic host membranes that contain macroscopically aligned proteorhodopsin species are expected to generate bulk ion gradients across host materials under illumination. Here, we present a solution-based synthetic protocol that allows high concentrations (up to 15 wt%) of active orientationally ordered proteorhodopsin species to be incorporated within nanostructured silica membrane hosts. Synthesis conditions and compositions were selected to stabilize proteorhodopsin molecules in the presence of the structure-directing surfactant and soluble network-forming silica species that co-assemble to form nanostructured silica host matrices, as established by small-angle X-ray diffraction analyses. Multidimensional solid-state NMR spectra show that proteorhodopsin molecules incorporated within nanostructured silica hosts retain native-like structures, though with some interesting differences. The optical absorbance behaviors of proteorhodopsin within the synthetic hosts are analogous to the photochemical reaction cycle of proteorhodopsin in native-like environments associated with the active transport of H^+-ions. The synthesis protocol is expected to be general and has been adapted to incorporate other functionally active membrane proteins within nanostructured silica host membranes. The versatile nanostructured silica-surfactant host materials open opportunities to integrate membrane proteins and their diverse functionalities into synthetic semi-permeable membranes.

461940 Suppression of Infrared Absorption in Nanostructured Metals By Controlling Faraday Inductance and Electron Path Length

Abstract:

While metals are widely used for electrodes in many optoelectronic devices, the device efficiency is limited by optical losses due to metals. As thinner devices are desired for reduced cost without sacrificing performance, the metal electrodes can be nanostructured to concentrate light in the photoactive material. A popular approach for the light concentration is to excite surface plasmon polaritons (SPPs). However, in such SPP-based devices, strong metal absorption is a major source of losses. The problem of this parasitic absorption becomes critical when a high device efficiency is desired. Moreover, if the device operates over a broad band as in solar cells, metal absorption needs to be controlled over the whole spectrum of interest and this poses a significant scientific and engineering challenge. Here, we investigate the loss suppression in metal nanocoil structures in the infrared range. Surprisingly, we find that the nanostructured metals can behave almost like a vacuum with a negligible optical loss in the infrared region. The loss is reduced in comparison to flat films by more than an order of magnitude over most of the very broad spectrum between short and long wavelength infrared. Our detailed analysis shows that this suppression of optical losses is due to both a large Faraday inductance and a long electron path. These two factors render the conduction electron in the metal effectively heavier than a hydrogen molecule. To investigate the usefulness of this effect in infrared detectors, we include a photoactive material in the coiled metal structures. In this case, the fraction of absorption in the active material increases by two orders of magnitude compared to non-coiled structures. Moreover, we demonstrate that the coiling of the infrared detector nanostrips increases the photoresponsivity by 15 times without additional electrical losses. These findings could benefit many metal-based applications that require low loss such as photovoltaics, photoconductive detectors, solar selective surfaces, infrared-transparent defrosting windows, and other metamaterials.

454155 Light Harvesting in Dye Sensitized Solar Cell Based on Consensitizer in Core-Shell Nanofiber Configuration Reducing Charge Recombination

Abstract:

Dye sensitized solar cell offers the potential of being a low-cost, high-efficiency photovoltaic device. However, the power conversion efficiency is limited as it cannot utilize all photons of the visible solar spectrum. A novel design of a core-shell photoanode is presented herein where a thin-shell of infrared dye is deposited on the core of sensitized TiO_2 nanofiber. Specifically, the ruthenium based dye (N719) sensitized TiO_2 nanofiber is wrapped by a thin-shell of copper phthalocyanine (CuPc). In addition to broadening the absorption spectrum, this core-shell configuration further suppresses the electron-hole recombination process. Instead of adopting the typical Förster resonance energy transfer, upon photons being absorbed by the infrared dye, electrons are transferred efficiently through a cascade process from the CuPc to the N719 dyes, the conduction band of TiO_2, the FTO electrode and finally the external circuit. This results in a high power conversion efficiency of 9.48% for the device. The proposed strategy provides an alternative method for enhancing the performance of DSSC for low-cost renewable energy in the future.

474206 Nanorod-like CH3NH3PbI3 for Planar Heterojunction Perovskite Solar Cell with Improved Performance

Abstract:

Hybrid organic-inorganic perovskite solar cell has recently attracted significant attention owing to the rapid improvement of the power conversion efficiency (PCE) as well as low cost manufacturing process. Great efforts involving high-quality perovskite film formation methodologies, compositional engineering of perovskite materials, and device architecture together with interface engineering have been made over the last six years, leading to a PCE leap from 3.8% in 2009 to the current world record of 22.1% in 2016 (NREL chart, http://www.nrel.gov/ncpv/images/efficiency_chart.jpg, Accessed 13.03.2016, 2016). The photo-to-electron conversion in perovskite solar cell involves several key steps, including light absorption by perovskite layer, charge transport in perovskite capping layer, and charge separation at n-type oxide/perovskite and perovskite/HTL interfaces. To achieve high photovoltaic performance, the morphology of perovskite is critically vital particularly for the plannar structured PSC, where the crystalline perovskite layer has to be self-standing without the support of a mesoporous scaffold.

This work reports a facile one-step HMT-assisted solution deposition method for reproducible fabrication of a high-quality nanorod-like $CH_3NH_3PbI_3$ thin films via controling over the $CH_3NH_3PbI_3$ crystallization in the presence of ammonia agent. SEM analysis review that a dense and grain-boundary-free perovskite capping layer can be prepared with orienting alignment of perovskite grains that fully cover the ammonia-modified ZnO substrate. XRD results demonstrate a high purity of $CH_3NH_3PbI_3$ thin film without no PbI_2 impurity. The perovskite film produces an appreciable enhancement in both photovoltaic performance and reproducibility as comparison with pristine perovskite, leading to an average PCE of 14.87±0.78%. The introduction of ammonia agent as nucleation-controlling and crosslinking agents and interface modifier provides a facile and effective approach for processing high purity and grain-boundary-free metal-halide perovskite thin film and tune the oxide/perovskite interface. The simplicity, ammonia agent-assisted and low temperature solution method can extend to construction of other perovskite-based hybrid optoelectronic devices.

473102 Lead-Free, Hybrid, Organic-Inorganic Halide for Light Harvesting

Abstract:

We will present the synthesis and optoelectronic properties of a lead-free, organic inorganic halide structures and show their integration as light harvesters in solar cells. Organic inorganic lead halide perovskite based solar cells have gained attention as a promising contender with nearly 20 % power conversion efficiencies. However, the environmental liabilities with the use and disposal of lead has led to the development of several non-lead based hybrid halides. The natural choice of many researchers has been tin (Group 14 , period 5) and to some extent bismuth (group 15, period 6). Although relatively new, the instability of tinâ€™s 2+ oxidation state has been a concern in spite of good semiconducting behavior. Here we will report the successful synthesis of a Ti based semiconductor with tunable bandgap based on the choice of the organic moiety. We will present and discuss the optical characteristics (ranging from 1.5 to 2.06) as well as the I-V characteristics of the of the various structures. While some of the synthesized products were clearly perovskites (based on XRD data), there are couple of structures whose crystalline structures have not yet been identified. The authors believe the XRD patterns to be a result of layered structures and expect to have it deciphered to present. The response of solar cell containing the hybrid halides coated onto a TiO_2 or ZnO will be presented.

463784 Improving Electron Transport in Nanostructured TiO_2 Electrode

Abstract:

Titanium dioxide (TiO_2) is one of the most widely used semiconductors in photovoltaics and photocatalysis because it is nontoxic, abundant, stable and photoactive. However, the wide bandgap, low electron mobility and short minority carrier diffusion length of TiO_2 limit its quantum efficiency in these applications. In this work, we present a solution chemical approach for making TiO_2 nanostructures for improving the electron transport in nanostructured TiO_2 electrodes.

461988 Symmetry-Breaking in Light-Trapping Nanostructures on Silicon for Solar Photovoltaics

Abstract:

While various materials have been investigated for photovoltaics, solar cells based on crystalline silicon (c-Si) dominate the current photovoltaics market. To reduce the cost of c-Si cells, wafer manufacturing companies have produced competitively priced thin c-Si films, ranging from a few microns to tens of microns, using a kerfless process. In such thin-film c-Si cells, light absorption becomes poorer than in thick films and light trapping is crucial to increase the photovoltaic efficiency. Han et al. have demonstrated that, among various light-trapping schemes, symmetry breaking in photonic nanostructures can approach the Lambertian light-trapping limit very closely. However, fabricating symmetry-breaking nanostructures in a scalable, cost-effective, manufacturable manner remains elusive. Here, we introduce a new approach to systematically break the symmetry in photonic nanostructures on c-Si surface. Using our approach, we fabricate low-symmetry inverted nanopyramid structures. Our method makes use of low-cost, manufacturable wet etching steps on c-Si(100) wafers without relying on expensive off-cut wafers. Our experiment and computational modeling demonstrate that the symmetry breaking can increase the Shockley-Queisser efficiency from 27.0 to 27.9% for a 10-micron-thick c-Si film. Further, our computation reveals that this improvement would increase from 28.1 to 30.0% with over-etching for a 20-micron-thick c-Si film.

464829 Computational and Kinetic Considerations for Morphology Prediction of Donor-Acceptor Oligomers for Organic Photovoltaics

Abstract:

Organic photovoltaic (OPV) solar cells are inexpensive, flexible, lightweight devices for sustainably generating electrical power. The efficiency and stability of OPV devices depends on their morphology, which we currently have poor synthetic control over. In this work, we perform molecular dynamics simulations of neat benzodithiophene-theinopyrrolodione (BDT-TPD) oligomers used in OPVs to evaluate the degree to which current simulation techniques are capable of predicting morphology in BDT-TPD films. The morphologies of simulated BDT-TPD volumes are characterized with simulated grazing-incidence X-ray scattering (GIXS) experiments and compared against experimental GIXS measurements.Â We find that the morphologies predicted from instantaneously quenched simulations better match experiments than structures that are allowed to equilibrate. This result highlights the competition between thermodynamics and kinetics in the structural evolution of polymer thin films.Â Further, we find that approximating the aromatic rings in BDT-TPD with rigid bodies rather than combinations of bond, angle, and dihedral constraints, results in better computational efficiency but poorer agreement with experimental structure. We determine the primary structural features in films of BDT-TPD are the stacking of oligomer backbones and the self-organization of backbone stacks into â€œribbonsâ€' with long-range spatial correlations. We conclude with a discussion of the shortcomings of current modeling tools and potential directions for improvement.

469339 Enhancing Dye Sensitized Solar Cell J-V Behavior By Integrating Nanoscale Polymer Films

Abstract:

Dye sensitized solar cells (DSSCs) are inexpensive, are easy to fabricate, have tunable optical properties such as color and transparency, and can be integrated into buildings and apparel easily. A DSSC is composed of three main components: (1) a photosensitive dye which converts sunlight into electrons, (2) a mesoporous TiO_2 photoanode which provides a large surface area for dye adsorption and a pathway for electron transport, and (3) a liquid electrolyte comprised of an iodide-triiodide redox couple which is used to regenerate the dye and acts as a hole transport medium. A major limitation of the current DSSC design is the liquid electrolyte, which is prone to leakage and evaporation along with being corrosive to metal contacts such as silver. Furthermore, electron recombination loss is significant at the TiO_2-electrolyte interface. Polymer electrolyte DSSCs can overcome these disadvantages and can potentially enhance the cell's photocurrent-voltage behavior [1]. However, due to the nanostructured, mesoporous nature of the TiO_2 photoanode and the effects of viscosity on fluid transport in such a structure, conventional polymer deposition techniques such as spin casting and dip coating lead to ineffective pore penetration. Often only the top 2 μm out of a >10 μm photoanode layer is in contact with the polymer with the remaining photoanode below completely void of polymer. This severely reduces device performance because a large portion of the adsorbed dye is not accessible and cannot be regenerated. To ensure optimal performing DSSCs, there needs to be intimate contact between the polymer electrolyte and the TiO_2 throughout the entire photoanode layer to ensure the redox couple can reach the adsorbed dye.

To overcome the limitations of liquid methods for incorporating polymer electrolytes into the mesoporous TiO_2, we employ a novel polymerization technique called initiated chemical vapor deposition (iCVD). iCVD is a solvent-free polymerization method that relies on gas-to-surface reactions at low pressures in which the reagents for polymerization, the monomer and initiator, are heated to vapors that can easily penetrate into the mesoscale voids of

the TiO

occurs conformally inside the pore surfaces. This makes it especially useful for cases that require penetration of sub-micron and nanometer-sized pores as issues of wettability and surface tension with liquid processing are absent. In addition, the lack of any solvent medium removes the possibility of liquids getting trapped in the photoanode, which may degrade performance, and are difficult to remove [2].

In this work, iCVD is used for the synthesis and integration of polymer electrolyte thin films within the mesoporous TiO_2 photoanode of DSSCs. Fourier transform infrared spectroscopy (FTIR), gel permeation chromatography (GPC), and X-ray photoelectron spectroscopy (XPS) analyses confirm that the polymers formed are stoichiometric in composition expected of linear homopolymers and chemically identical to ones formed with liquid solution methods. The surface reaction kinetics and mass transport have been found to depend on the fractional surface saturation of monomer, , which is an adsorption parameter that provides a measure of the surface availability of the monomer. By carefully controlling , conformal coatings throughout the TiO_2 photoanode have been achieved. To gain a better understanding on the effect of the polymer electrolyte on DSSC behavior, current-voltage, UV-VIS absorption, incident photon-to-current efficiency (IPCE) and electrochemical impedance spectroscopy (EIS) measurements have been employed. Experimental results indicate that DSSCs with ultrathin polymer electrolyte thin films integrated on the photoanode pore surface can enhance cell performance. For example, by applying poly(1-vinylimidazole) (PVIZ), power conversion efficiencies are found to be 27% higher than liquid-electrolyte DSSCs due to a simultaneous increase in short circuit current J_{sc} and open circuit voltage V_{oc} [3]. To our knowledge, this is the first reported demonstration of PVIZ as a polymer electrolyte in DSSCs.

1. Smolin, Y. Y.; Nejati, S.; Bavarian, M.; Lee, D.; Lau, K. K. S.; Soroush, M. *Journal of Power Sources* **2015,** 274, 156-164.

2. Nejati, S.; Lau, K. K. *Nano letters* **2010,** 11, (2), 419-423.

3. Kuba*, A. G.; Smolin*, Y. Y.; Soroush, M.; Lau, K. K. S. *Chemical Engineering Science. Special Issue: Energy Conversion and Storage* **2016,** (*in press.*) http://dx.doi.org/10.1016/j.ces.2016.05.007 (*equal contribution)

468024 Photocatalytic Photosystem I/PEDOT Composite Films Prepared By Vapor Phase Polymerization

Abstract:

Photosystem I (PSI) is an integral membrane protein complex contained within the chloroplast granum of plants and cyanobacteria. The product of over a billion years of systematic evolution, PSI is adapted to the earth's climate and sun's radiant energy profile. PSI transduces blue-violet and red light, sustaining 1.1eV of stored energy in <100ps by way of a cascade of irreversible internal reactions. Internal quantum efficiency for this process is nearly 100%, making PSI a model photodiode. As such, PSI has garnered interest as a macromolecule for use in photovoltaic and photocatalytic systems in a variety of roles. Recently, there have been efforts to include PSI within composite films, whereby the high potential and photocatalytic efficiency of PSI is coupled with an electrically conductive matrix.

We describe a versatile technique to fabricate photocatalytic composite films containing PSI and a chosen intrinsically conductive polymer (ICP). Iron (III)—a friedel crafts catalyst—is added to protein solutions by way of water soluble $FeCl_3$. Drop casting of these solutions on a substrate of choice is followed by contact with low temperature (40°C) monomer vapor of a well-studied ICP, poly(3,4-etylenedioxythiophene) PEDOT. In minutes, a ~200 nm thick polymer/protein composite is grown. Composite films retain the high conductivity of the used ICP and the characteristic absorbance and photocatalytic capabilities of PSI multilayers. Vapor phase processing allows versatility in choice of underlying substrate.

466440 Computational Investigations of Drug Storage and Delivery in Bio-Compatible Nanoporous Materials

Abstract:

Encapsulation of drug molecules is required to improve the drug bio-distribution, the biological half-life of the active species and therapeutic effect of the drug.[1] Lipids, polymeric nanoparticles, metal clusters and carbon structures have been tested as drug storage materials but these materials have several limitations such as low drug loading capacities (<5 wt%), rapid drug release and toxicity.[2] Recently, metal organic frameworks (MOFs), a new group of nanoporous materials, have been studied in drug and cosmetic molecule storage applications due to their large surface areas, high pore volumes and tunable pore sizes and several MOFs are identified as promising materials for drug storage.[3] However, considering the very large number of available MOFs, it is not practical to assess the potential of each MOF in drug storage applications using purely experimental manners. Computational methods that can rapidly predict drug storage capacity of MOFs prior to experimental studies would be highly useful to identify the materials with the highest performances. However, there is almost no information about the drug storage performances of different types of MOFs. This lack of information also limits understanding the relation between drug storage performances of MOFs and their structural properties. In this work, we studied both adsorption and diffusion of representative drug and cosmetic molecules in various bio-compatible MOFs. Ibuprofen (an analgesic and anti-inflammatory drug) was chosen as a representative drug molecule, caffeine (lipo-reducer) and urea (hydrating agent) were considered as representatives of cosmetic molecules. Ibuprofen, caffeine and urea storage performances of 24 different bio-compatible MOFs were examined using configurational bias Monte Carlo (CBMC) simulations. Once a drug molecule is captured by a MOF, it has to be delivered to the target cells and released over a long period of time to increase its therapeutic effect. Short delivery time is not desired since in this case drug release occurs prior to reach to targeted cells. Therefore, release performance of materials must be also examined in order to identify the best performing materials as drug carriers. With this aim, we performed molecular dynamics simulations to examine ibuprofen, caffeine and urea diffusion in a MOF that showed high storage performance. Both rigid and flexible MOF structures were used in diffusion simulations to understand the effect of structure flexibility on the drug/cosmetic molecule diffusion. We also studied the effect of water on diffusion of drug and cosmetic molecules since these molecules are generally encapsulated in a liquid environment in practical applications. Finally, effects of MOFsâ€™ structural properties such as pore volume and pore size on the drug storage and diffusion were investigated to provide information on the materialsâ€™ structure-performance relationships.[4] This study showed that bio-compatible MOFs have strong potential in drug storage and delivery and they can be alternatives to traditional drug carriers. The computational methodology that we described here will be helpful to identify the most promising MOFs among many of them for drug/cosmetic storage prior to extensive experimental efforts. **Acknowledgment:** Financial supports provided by the TUBITAK 2211-C Scholarship Program. S.K. acknowledges TUBA-GEBIP Programme. *Corresponding author: skeskin@ku.edu.tr **References:** [1] C. Tamames-Tabar *et al.*, MOFs in Pharmaceutical Technology. In Bio- and Bioinspired Nanomaterials, Wiley-VCH Verlag GmbH & Co. KGaA: Germany, 2014. [2] P. Horcajada *et al.*, Chem. Rev. **112**, No. 2, 1232 (2011). [3] G. Férey *et al.*, Science, **309**, No. 5743, 2040 (2005). [4] I. Erucar and S. Keskin, Ind. Eng. Chem. Res., **55**, No. 7, 1929 (2016).

471914 Nanoharvesting of Polyphenolic Flavonoids from *Solidago Nemoralis* Hairy Root Cultures Using Functionalized Mesoporous Silica Nanoparticles

Abstract:

Plant secondary metabolites are produced in response to the environment of certain plants as a defense mechanism against predators and other stressors. Sometimes, these metabolites also bind to human receptors to induce favorable responses to diseases and pathogens, which can be enriched in a particular plant cell cultures by genetic and environmental manipulations. The conventional method of recovery of natural products from these genetically engineered plants is to grow the plants to a certain degree of maturity and then to harvest whole tissues, macerate the tissues to gain access to the compounds, and extract them. Nanoharvesting offers an alternative for continuous production of secondary metabolites from plant cell cultures, in which nanoparticles are designed to bind and carry molecules out of living cells. The carriers are to be designed so that they enter into plant cells and are released without inducing significant toxicity, and with specific binding of compounds of interest.

Here, nanoharvesting of polyphenolic flavonoids, enriched in genetically modified *S. nemoralis* hairy root cultures, is carried out using mesoporous silica nanoparticles (MSNPs) functionalized with both amines and titanium dioxide (TiO_2). Amine functionalization was performed to facilitate uptake of nanoparticles into plant cells, and TiO_2 functionalization to provide coordination binding sites for metabolites. Non-magnetic MSNPs (170 nm diameter) and magnetic MSNPs with Fe_3O_4 cores (70 nm diameter) with highly porous structures were synthesized and functionalized with amines and TiO_2. Both particles were shown to be taken up in *S. nemoralis* hairy roots, and recovered using external centrifugal and magnetic forces, respectively. Intracellular uptake and localization of the nanoparticles (0-1000 μg/ml in Murashige and Skoog media) in hairy roots were visualized by fluorescent imaging, after tagging the nanoparticles with rhodamine B isothiocyanate. Quenching of fluorescence in bulk solution using trypan blue was used to confirm intracellular localization of the tagged particles. Post-uptake viability of hairy roots was demonstrated by a concentration-dependent growth study. Proof of the flavonoid nanoharvesting concept was inferred from the color change (white to orange) of the recovered nanoparticles after exposure to hairy roots. Quantitative determination and analysis of nanoharvested flavonoids will be reported based on antiradical activity of metabolites bound to particles and HPLC.

451087 Modeling: A Tool for Experimentalists. Design, Synthesis and Evaluation of Self-Assembling Dendrons for Gene/Drug Delivery

Abstract:

Due to the relative easy synthesis and commercial availability, nanovectors based on self-assembling units are among the most utilized non-viral vectors for gene/drug transfer. Contextually, recent advances in molecular simulations and computer architectures allow for accurate predictions of many structural, energetic, and eventual self-assembly features of these nanocarriers *per se* and in complex with their genetic cargoes. In the present work, we aim at reviewing our own, most recent efforts in the field of computer-assisted design, synthesis, *in vitro*and in vivo activity of these (nano)biomaterials [1-9]. Specifically, we will discuss two cases:

1) Anticancer drug nanomicelles formed by self-assembling amphiphilic dendrimer able to overcome drug resistance in cancer cell [2]

Drug resistance and toxicity constitute challenging hurdles for cancer therapy. The application of nanotechnology for anticancer drug delivery is expected to address these issues and bring new hope for cancer treatment. In this context, we established an original nanomicellar drug delivery system based on an amphiphilic dendrimer (*AmDM*), which could generate supramolecular micelles to effectively encapsulate the anticancer drug doxorubicin (DOX) with high drug-loading capacity (>40%). The resulting *AmDM*/DOX nanomicelles were able to enhance drug potency and contrast doxorubicin resistance in breast cancer models by significantly enhancing cellular uptake while considerably decreasing efflux of the drug. In addition, the *AmDM*/DOX nanoparticles abolished significantly the toxicity related to the free drug. Collectively, our studies demonstrate that the drug delivery system based on nanomicelles formed with the self-assembling amphiphilic dendrimer constitutes a promising and effective drug carrier in cancer therapy.

2) Chameleon self-assembling nanovectors for efficient siRNA delivery *in vitro* and *in vivo* [4]
Small interfering RNA (siRNA) delivery remains a major challenge in RNAi-based therapy. Through the combined action of *in silico*/experimental efforts, we designed and produced for the first time an amphiphilic dendrimer *AD* able to self-assemble into adaptive supramolecular assemblies upon interaction with siRNA. The resulting nanovectors can effectively deliver siRNAs to various cell lines, including human primary and stem cells, thereby outperforming the currently available non-viral vectors. Most importantly, however, *AD*-mediated gene silencing was successfully achieved also *in vivo*.

References

[1] P. Posocco, E. Laurini, V. Dal Col, D. Marson, K. Karatasos, M. Fermeglia and S. Pricl, "Tell me something that I do not know. Multiscale molecular modeling of dendrimer/dendron organization and self-assembly for gene therapy," *Curr. Med. Chem.*, vol. 19, pp. 5062-5087, 2012.

[2] T. Wei, C. Chen, J. Liu, C. Liu, P. Posocco, X. Liu Q. Cheng, S. Huo, Z. Liang, M. Fermeglia, S. Pricl, X. J.Liang, P. Rocchi, L. Peng, "Anticancer drug nanomicelles formed by self-assembling amphiphilic dendrimer to combat cancer drug resistance," *Proc. Natl. Acad. Sci. USA*, vol. 112, pp. 2978-2983. 2015.

[3] D. Marson, E. Laurini, P. Posocco, M. Fermeglia, S. Pricl, "Cationic carbosilane dendrimers and oligonucleotide binding: an energetic affair," *Nanoscale*, 7, pp. 3876-87, 2015.

[4] X. Liu, J. Zhou, T. Yu, C. Chen, Q. Cheng, K. Sengupta, Y. Huang, H. Li, C. Liu, Y. Wang, P. Posocco, M. Wang, Q. Cui, S. Giorgio, M. Fermeglia, F. Qu, S. Pricl, Y. Shi, Z. Liang, P. Rocchi, J. J. Rossi, L. Peng, "Adaptive amphiphilic dendrimer-based nanoassemblies as robust and versatile siRNA delivery systems," *Angew. Chem. Int. Ed. Engl.*, vol. 53, pp. 11822-11827, 2014.

[5] S. Kala, A. S. Mak, X. Liu, P. Posocco, S. Pricl, L. Peng, A. S. Wong, "Combination of dendrimer-nanovector-mediated small interfering RNA delivery to target Akt with the clinical anticancer drug paclitaxel for effective and potent anticancer activity in treating ovarian cancer," *J. Med. Chem.*, vol. 57, pp. 2634-42, 2014.

[6] P. Posocco, X. Liu, E. Laurini, D. Marson, C. Chen, C. Liu, M. Fermeglia, P. Rocchi, S. Pricl, L. Peng, "Impact of siRNA overhangs for dendrimer-mediated siRNA delivery and gene silencing," *Mol. Pharm.*, vol. 10, pp. 3262-3273, 2013.

[7] X. Liu, C. Liu, E. Laurini, P. Posocco, S. Pricl, F. Qu, P. Rocchi, L. Peng, "Efficient delivery of sticky siRNA and potent gene silencing in a prostate cancer model using a generation 5 triethanolamine-core PAMAM dendrimer," *Mol. Pharm.*, vol. 9, 470-81, 2012.

[8] X. Liu, J. Wu, M. Yammine, J. Zhou, P. Posocco, S. Viel, C. Liu, F. Ziarelli, M. Fermeglia, S. Pricl, G. Victorero, C. Nguyen, P. Erbacher, J. P. Behr, L. Peng, "Structurally flexible triethanolamine core PAMAM dendrimers are effective nanovectors for DNA transfection in vitro and in vivo to the mouse thymus," *Bioconjug. Chem.*, vol. 22, 2461-2473, 2011.

[9] A. Barnard, P. Posocco, S. Pricl, M. Calderon, R. Haag, M. E. Hwang, V. M. Shum, D. W. Pack, D. K. Smith, "Degradable self-assembling dendrons for gene delivery: experimental and theoretical insights into the barriers to cellular uptake," *J. Am. Chem. Soc.*, vol. 133, pp. 20288-20300, 2011.

457771 Investigation of CNT-Induced Bacteria Lysis and Protein Release

Abstract:

Carbon nanotubes (CNTs) were investigated for their potential use as lysis agents for bacterial cells. Recombinant *Escherichia coli* BL21 (DE3) expressing b-lactamase and Green Fluorescent Protein (GFP) were exposed to various concentrations of CNTs under various levels of agitation and exposure times. Fluorescence assay for GFP, enzymatic activity for b -lactamase activity, and Transmission Electron Microscopy (TEM) were used to characterize the extent of lysis. A correlation was also developed that related the extent of protein release to the concentration of CNTs and agitation rate. The correlation indicated that the agitation rate and CNT concentration affect lysis in a similar fashion.

459693 Using Fluoroalkylated Polyethylene Glycol-Stabilized Perfluorocarbon Nanodroplet As Oxygen Carriers to Reduce Oxygen Inhibition Effect for Microalgal Growth

Abstract:

Introduction: Dissolved oxygen (DO) built up in the photobioreactor and/or its excessive accumulation in the medium have long been considered as one of the major limitations for closed and massive cultivation of microalgae. Although the DO may be removed via air sparging-based photobioreactor technology, disadvantages such as high cost, increased contamination, and severe cell damage caused by hydrodynamic stress remain obstacles for microalgal culture. Perfluorocarbon (PFC), a fluorine-substituted derivative of hydrocarbons which can dissolve large respiratory and other non-polar gases as compared to water, has been widely used as an oxygen/carbon dioxide transporter in various fields. To overcome the effect of oxygen inhibition on the microalgal growth, a strategy of using perfluorocarbon nanodroplets (PNDs) as oxygen scavengers was explored.

Materials and Methods: The PNDs were synthesized by perfluorooctyl bromide (PFOB) and fluoroalkylated methoxy polyethylene glycol (R_F-MPEG, Mw of PEG molecule = 2000 or 5000) with various formulations through single PFC-in-water emulsification in association with solvent evaporation approach and were characterized using Fourier Transform Infrared spectroscopy. The stability of each type of PND under dynamic environment was evaluated by analyzing their changes in size and concentration while circulating in the photobioreactor system for 7 days. The deoxygenability of each type of PND under 7-days oxygenated photobioreactor operation was determined based on the mass accumulation rate coefficient (K_La^{ac}) of DO in the medium by use of dynamic method. To verify the effectiveness of PNDs-aided photobioreactor on *Nannochloropsis Oculata (N. oculata)*growth under a hypercritical oxygen environment, the cellular growth rate in each group was examined through measurement of the cell concentrations by hemocytometry every 24 h for up to 7 days.

Results and Discussion: Our results showed that all of the PNDs expanded during circulation in the photobioreactor and the one prepared by R_F-MPEG-2000 (MW of PEG = 2000) and perfluorooctyl bromide (PFOB) with a volume ratio of 9:1 (PNDs-2000/PFOB (9:1)) exhibited the lowest size increase of 19% (from 200 ± 2.43 to 238 ± 1.99 nm) after 7 days. Moreover, the concentration of DO in the medium approximately decreased 25%, 15%, 9%, and 5% at day 7 where the value of K_La^{ac} was about 4.7, 5.92, 6.49, and 7.4 $\times 10^{-3}$ h^{-1} for the group with 1% (v/v) PNDs-2000/PFOB (9:1), PNDs-2000/PFOB (7:3), PNDs-5000/PFOB (9:1), and PNDs-5000/PFOB (7:3), respectively, as compared to the group without PND. These results clearly exhibited that the PNDs-2000/PFOB (9:1) can provide the highest stability and capacity of deoxygenation for the defined photobioreactor operation. Under oxygenated cultivation, the growth of *N. oculata* (μ) without PNDs was tremendously arrested (μ = 0.366 day^{-1}) within seven days, while the cells with aid of 1% (v/v) PNDs-2000/PFOB (9:1) successfully multiplied in the hyperoxygenated environment that the growth rate of 0.467 day^{-1} was comparable to the normal breeding with neither PNDs nor oxygen injection (μ = 0.498 day^{-1}), and the biomass production was five-fold higher than the O_2-treated group without PND after cultivation for seven days.

Conclusions: In summary, the developed PNDs-2000/PFOB (9:1) enabled to efficiently remove the DO in the photobioreactor that provided a feasible means for massive culture of microalgae in closed setting.

465526 The Influence of Surface Functionalization on Nanoparticle-Cellular Interactions

Abstract:

Owing to its exceptional antimicrobial activity silver nanoparticles (AgNPs) warrant the significant incorporation in the consumer and medical products to date. However, an important aspect when considering the use of AgNPs in for biomedical applications is the interaction of cells and proteins with nanoparticles. Proteins interact with NPs almost instantly, forming a protein-corona (PC) at the surface of the NPs. This PC remains associated with the NPs under in vivo conditions, thereby conferring the biological identity to the particles. Studies have been conducted on the interaction of NPs with human serum proteins, but there is a lack of information on the fate of these serum-coated NPs and how they interact with the proteins at the cellular level. The study aims to investigate the role of bioactive compounds (Curcumin and EGCG) in the AgNP-protein corona formation and its further interactions with cells. In my presentation I will specifically present the data on synthesis and characterisation of Curcumin and EGCG coated nanoparticles and discuss how these coated nanoparticles interact differently with cells and protein milieu. Further applicability of these bioactive coated nanoparticles for biomedical applications will be discussed.

468114 Synthesis and Characterization of Pegylated Self-Assembled Rosette Nanotubes As Drug Delivery Vehicles

Abstract:

Rosette nanotubes (RNTs), which are synthetic hybrid DNA nucleic-acid nanotubes, are self-assembled from fused guanine and cytosine motifs (GÎ›C motif). In RNTs, DNA Watson-Crick pairing takes place between the guanine side of the GÎ›C motif and its cytosine side. The spatial arrangement of these arrays leads them to form a six-membered supermacrocycle with high stability as a result of the extensive H-bond network and the inter-rosette Ï€-Ï€ stacking interactions. In addition, using synthetic approaches, varied functional groups have been grafted to GÃ™C such as amino acid, peptides, and crown ethers. Upon self-assembly, the functional groups will display on the surface of RNTs which imparted important physical and chemical properties.

RNTs have found numerous medical applications as a result of their high biocompatibility and low toxicity. In order to improve the solubility in aqueous and pharmacokinetics behavior of RNTs as drug delivery vehicles, we have developed polyethylene glycol (PEG) covalently grafted on the surface of RNTs through organic synthesis and self-assembly, in which PEG is non-toxic and widely used to improve chemical and physical properties of drug delivery vehicles. As a result, the surfaces of RNTs have been grafted PEG chains. Here we will present our synthetic strategy and self-assembly studies of this new material as well as *in vitro* cytocompativility studies.

461158 Photoexcited Quantum Dots for Killing Multidrug-Resistant Bacteria

Abstract:

Multidrug-resistant bacterial infections are an ever-growing threat because of the shrinking arsenal of efficacious antibiotics. Metal nanoparticles can induce cell death, yet the toxicity effect is typically nonspecific. Here, we show that photoexcited quantum dots (QDs) can kill a wide range of multidrug-resistant bacterial clinical isolates, including methicillin-resistant Staphylococcus aureus, carbapenem-resistant Escherichia coli, and extended-spectrum Î²-lactamase-producingÂ Klebsiella pneumoniaeÂ and Salmonella typhimurium. The killing effect is independent of material and controlled by the redox potentials of the photogenerated charge carriers, which selectively alter the cellular redox state. We also show that the QDs can be tailored to kill 92% of bacterial cells in a monoculture, and in a co-culture of E. coli and HEK 293T cells, while leaving the mammalian cells intact, or to increase bacterial proliferation. Photoexcited QDs could be used in the study of the effect of redox states on living systems, and lead to clinical phototherapy for the treatment of infections.

470547 Chitosan-Coated Selenium Nanoparticles and Their Affects on Bacterial Growth Kinetics

Abstract:

Introduction:

The prevalence of antibiotic-resistant bacteria has continued to grow in recent years, resulting from increased use (and misuse) of antibiotics in daily life. It is imperative, then, to develop novel bactericidal treatments and therapies to combat these pathogens. Unfortunately, the discovery of new antibiotics, in the traditional sense, has become increasingly difficult in recent years. Selenium is an essential trace material found in the body associated with antioxidant and metabolic mechanisms that play a role in preventing cell damage. Selenium nanoparticles (SeNP) have been previously found to be antibacterial without any apparent toxicity to healthy mammalian cells. Here, we have developed SeNP synthesized with a chitosan coating to improve colloidal stabilization and to introduce new surface chemistries for further targeting and delivery options. The chitosan-coated SeNP (Ch-SeNP) were able to conserve the size distribution and the antibacterial efficacy seen with previous iterations of SeNP. To compare these results, we employed the Gompertz kinetic growth models to analyze the changes in lag time (time in which it takes the bacteria to enter their exponential growth phase) and maximum growth rate, with respect to different concentrations of SeNP and Ch-SeNP. The results showed an increase in the lag time and a decrease in the growth of *S. aureus* bacteria with Ch-SeNP.

Materials and Methods:

Ch-SeNP were produced using a colloidal suspension method. Briefly, chitosan in a 1% acetic acid solution was added dropwise into a stirred sodium selenite solution. Ascorbic acid was then added dropwise to the solution and stirred until equilibrium was reached. The nanoparticles were centrifuged for 10 minutes to remove excess reactants, and

subsequently resuspended into DI water. DLS and TEM were used to characterize the Ch-SeNP to determine the size and surface charge. The growth kinetics were measured by incubating *S. aureus*with various Ch-SeNP concentrations in 3% TSB. The optical density (at 592 nm) was measured regularly for 24 hours throughout the growth assay. Gompertz and Richards growth models were applied to develop the growth kinetics to understand the role of the selenium nanoparticles on the bacteriaâ€™s proliferation. All experiments were replicated three times and statistical significance analyzed via ANOVA. The data obtained was compared with controls using a two-tailed, unpaired Studentâ€™s t-test, with $p<0.05$ considered to be statistically significant.

Results and Discussion:

The Ch-SeNP inhibited *S. aureus*growth over time. Most notably, the lag time of the growth curves increased while the growth rate decreased with increasing concentrations of Ch-SeNP treatments. Even for Ch-SeNP concentrations as low as 0.0023 mg/ml, both the lag time and growth rate changed significantly compared to untreated controls.

Conclusion:

Selenium nanoparticles coated with chitosan have easily reproducible batch-to-batch characteristics and are able to effectively inhibit the growth of *S. aureus*. Using the Gompertz growth model, the present study accurately calculated the growth kinetics of *S. aureus* with the chitosan coated selenium nanoparticles. Future studies can use the Gompertz and Richards model to accurately standardize the effectiveness of nanoparticles on the growth of *S. aureus*. Moreover, further studies regarding the mechanism of antibacterial action of selenium nanoparticles coated with chitosan are needed.

Acknowledgements:

The authors thank Northeastern University and the Department of Chemical Engineering for funding this research.

468698 Targeted Silver Nanoparticles for Selective Cytotoxicity of Tumorigenic Endothelial Cells

Abstract:

Silver nanoparticles (AgNPs) have widely documented cytotoxic properties that are influenced by size, shape, and surface chemistry. AgNPs introduced to a biological environment interact with proteins and cell membrane receptors, which influence cell uptake, gene expression and cytotoxicity. Thus, controlling the cell-NP interface has important implications in the application of AgNPs for therapeutic purposes. For example, AgNPs can be engineered to selectively interact with specific cell types by surface functionalization with biomimetic targeting moieties. The objective of this study was to investigate the effects of AgNP exposure on the cytotoxicity of hemangioendothelioma (EOMA) cells, which are a tumorigenic endothelial cell line. Cells were incubated with citrate stabilized AgNPs of variable diameter (10-100μm) at a 0.01 mg/mL. Absorbance based assays revealed that higher cell uptake and cytotoxic responses correlated with decreased particle size. Alternatively, AgNPs were functionalized with cyclic RGD peptides by either physisorption or covalent conjugation to target $\alpha_v\beta_3$-integrin receptors on the cell membrane of EOMA cells. The presence of $\alpha_v\beta_3$-integrin expression in EOMA cells was demonstrated via RT-PCR and confocal microscopy. EOMA cells exposed to RGD-AgNPs in serum-containing media did not show a significant enhancement in cell uptake or cytoxocity relative to non-binding RAD control peptides. In contrast, RGD-AgNP fabricated with NHS-ester reaction chemistry elicited a significant increase in cytotoxicity relative to non-targeted controls. DCFDA and ELISA assays suggest that enchanced cytotoxicity is correlated with increased ROS generation and an upregulation of inflammatory cytokine production. These results demonstrate the utility of functionalizing AgNP with RGD peptides for selective cytotoxicity of $\alpha_v\beta_3$-integrin expressing EOMA cells. Future work will be aimed at further elucidating the molecular mechanisms of RGD-AgNP induced cytotoxicity in EOMA cells.

473075 Cell Membrane-Camouflaged Nanomotors for Biodetoxification and Drug Delivery

Abstract:

Attempts to apply artificial micro-/nanomotors for diverse biomedical applications have inspired a variety of strategies for designing motors with diverse propulsion mechanisms and functions. However, existing artificial motors are made exclusively of synthetic materials, which are subject to serious immune attack and clearance upon entering the bloodstream. Herein we present a platform by using red blood cell as camouflage coating for micro-/nanomotors. As the first example, a red blood cell membrane-camouflaged nanowire that can serve as new generation of biomimetic motor sponge is described. The biomimetic motor sponge is constructed by the fusion of biocompatible gold nanowire motors and RBC nanovesicles. The motor sponge possesses a high coverage of RBC vesicles, which remain totally functional due to its exclusively oriented extracellular functional portion on the surfaces of motor sponge. These biomimetic motors display efficient acoustical propulsion, including controlled movement in undiluted whole blood. The RBC vesicles on the motor sponge remain highly stable during the propulsion process, conferring thus the ability to absorb membrane-damaging toxins and allowing the motor sponge to be used as efficient toxin decoys. The efficient propulsion of the motor sponges under an ultrasound field results in accelerated neutralization of the membrane-damaging toxins. Such motor sponges connect artificial nanomotors with biological entities and hold great promise for treating a variety of injuries and diseases caused by membrane-damaging toxins.
As another example, an elegant approach is used to directly turn natural red blood cells (RBCs) into functional micromotors with the aid of ultrasound propulsion and magnetic guidance. Since the RBC motors preserve the biological and structural features of regular RBCs, these motors possess a wide range of antigenic, transport, and mechanical properties that common synthetic motors cannot achieve and thus hold considerable promise for a number of practical biomedical uses. The ability of the RBC micromotors to transport imaging and therapeutic agents at high speed and spatial precision through a complex microchannel network is also demonstrated. Such ability to load and transport diagnostic imaging agents and therapeutic drugs within a single cell-based motor, in addition to a lower toxicity observed once the drug is encapsulated within the multicargo RBC motor, opens the door to the development of theranostic micromotors that may simultaneously treat and monitor diseases.

457102 Plasmon-Enhanced Photocatalytic CO_2 Reduction on Nanostructured Composite Electrodes

Abstract:

Using inputs of only sunlight, electricity, carbon dioxide, and water, photocatalytic CO_2 reduction could mitigate the effects of climate change caused by the burning of fossil fuels and produce valuable fuels or chemical feedstocks. However, current CO_2 reduction technologies suffer from high overpotentials and low selectivity, producing a mixture of carbon monoxide, methane, ethylene, formate, methanol, and other products.

Plasmon-assisted photocatalytic CO_2 reduction leads to greater selectivity and lower overpotentials by unlocking unique mechanistic pathways. It has been well-documented that the strongly localized near fields at the surface of plasmonic nanoparticles promote electron-hole pair generation in nearby semiconductors.[1–3] Plasmon decay generates an excited "hot" electron which can be transferred to a surface molecule for direct reduction or injected into an adjacent wide band gap catalyst, effectively limiting carrier recombination through charge separation and expanding the usable portion of the solar spectrum.[4–7] The electron dynamics in an irradiated plasmonic nanoparticle can alter the electronic coupling with surface adsorbed CO_2 and reaction intermediates, thereby changing the binding energy of these species and the catalytic properties of the plasmonic metals. This plasmon-enhanced effect was demonstrated when optically excited surface plasmons at the roughened surface of an Ag electrode in CO_2-containing solutions produced strong enhancement of the cathodic current.[8–9] The wide tunability of the plasmon resonance frequency with shape, size, and material confers fine control over these catalytic mechanisms, allowing for optimization of the photocatalytic performance.

To take advantage of plasmonic catalysis, nano-sized spheres, cubes, and wires of gold, silver, and copper

have been synthesized as these materials exhibit strong plasmonic and electrocatalytic behavior. For the semiconducting component of the photoelectrodes, nanostructured TiO_2 co-catalysts have also been prepared. Transmission electron microscopy (TEM) and scanning electron microscopy (SEM) were used to determine the shape and size distribution of the nanoparticles. X-ray diffraction (XRD) and optical spectroscopy were employed to characterize the electrodes. The CO_2 reduction ability of the plasmonic metal nanoparticles with and without a semiconductor co-catalyst has been tested in a custom glass photoelectrochemical cell with a quartz window that allows for front- and back-illumination of the electrode. The two electrode chambers are separated by a membrane and can achieve a gas-tight seal that allows for precise pressure measurements. Experiments were run both under isolation of the pressurized cell and with a constant CO
spectrometer (GC-MS) that allows detection and quantification of gaseous products. *Ex-situ* high performance liquid chromatography (HPLC) was used to quantify liquid products.

This material is based upon work performed by the Joint Center for Artificial Photosynthesis, a DOE Energy Innovation Hub, supported through the Office of Science of the U.S. Department of Energy under Award Number DE-SC0004993.

The work is supported by the National Science Foundation Graduate Research Fellowship under Grant No. 1106400.

Work at the Molecular Foundry was supported by the Office of Science, Office of Basic Energy Sciences, of the U.S. Department of Energy under Contract No. DE-AC02-05CH11231.

References:

(1) Thomann, I.; Pinaud, B. A.; Chen, Z.; Clemens, B. M.; Jaramillo, T. F.; Brongersma, M. L. *Nano Lett.* **2011**, *11*, 3440–3446.

(2) Ingram, D. B.; Linic, S. *J. Am. Chem. Soc.* **2011**, *133*, 5202–5205.

(3) Gao, H.; Liu, C.; Jeong, H. E.; Yang, P. *ACS Nano* **2012**, *6*, 234–240.

(4) Mukherjee, S.; Zhou, L.; Goodman, A. M.; Large, N.; Ayala-Orozco, C.; Zhang, Y.; Nordlander, P.; Halas, N. J. *J. Am. Chem. Soc.* **2014**, *136*, 64–67.

(5) Mubeen, S.; Lee, J.; Singh, N.; Krämer, S.; Stucky, G. D.; Moskovits, M. *Nat. Nanotechnol.* **2013**, *8*, 247–251.

(6) Naya, S.; Teranishi, M.; Isobe, T.; Tada, H. *Chem. Commun.* **2010**, *46*, 815–817.

(7) Lee, J.; Mubeen, S.; Ji, X.; Stucky, G. D.; Moskovits, M. *Nano Lett.* **2012**, *12*, 5014–5019.

(8) Kostecki, R.; Augustynski, J. *Chem. Phys. Lett.* **1992**, *194*, 386–390.

(9) Kostecki, R.; Augustynski, J. *J. Appl. Phys.* **1995**, *77*, 4701–4705.

459985 Photon, Electron, and Ion Management in Artificial Photosynthesis: Realizing Efficient Renewable Energy to Fuel Conversion

Abstract:

Direct production of fuels from sunlight, air (N_2 or CO_2), and water that can be stored, transported, and later converted into hydrogen or electricity to provide power for transportation and distributed energy generation, have received recent attentions worldwide. This technology could also provide chemicals as synthetic precursors or realize grid-level storage of intermittent solar energy. In artificial photosynthesis based on semiconductor photoelectrochemistry, production of chemical fuels generally requires the coupling of separated electrical charges with electrocatalysts for multi-electron chemical reactions. Meanwhile, the development of integrated, efficient and stable photoelectrochemical (PEC) systems requires the pairing of light-absorbing materials with an optimum bandgap combination. The development of such systems has been hindered in part by the lack of semiconducting materials that can simultaneously provide efficiency and stability in a corrosive aqueous electrolyte, typically strong acid or base. In this talk, I will present our recent progress in the development of efficient and stable PEC devices for H2 production from solar-driven water splitting. First, I will summarize some recent development of photon management in PEC devices and module designs from modeling to experiment. Then, I will discuss methods for heterogeneous interfacial energetic modification on covalence-bond semiconductors for efficient free carriers (electrons/holes) separation. Finally, I will present a novel approach to managing the ionic environment for creating a defect-tolerant condition, which leads to a record-setting solar water-splitting prototype.

462500 Tandem Core-Shell Si-Ta_3N_5 Photoanodes for Photoelectrochemical Water Oxidation

Abstract:

Tantalum nitride (Ta_3N_5

type doping, relatively small bandgap of 2.1 eV, and band positions that straddle the redox potentials of both the hydrogen and oxygen evolution reactions.[1] However, there are other challenges associated with Ta_3N_5 including poor charge transport properties, late photocurrent onset and photodegradation. To address all of these challenges, we fabricated efficient and stable core-shell heterostructured Si-Ta_3N_5 nanowire photoanodes coated with $CoTiO_x$ and NiO oxygen evolution co-catalysts.

Previous device architecture strategies for metal oxide semiconductors with charge transport limitations have included nanostructuring or scaffolding. In this work, a similar approach was applied to Ta_3N_5 to make a nanostructured core-shell Si-Ta_3N_5 device where the Si is a semiconducting, nanowire scaffold and the Ta_3N_5 is coated onto Si as a thin shell through which charge extraction should be efficient. With n-type Si as the scaffold, a heterojunction between Si and Ta_3N_5 is formed enabling a dual-absorber configuration resulting in a 200 mV cathodic shift for photocurrent onset due to the photovoltage gained from Si and therefore improves the overall performance of the Si-Ta_3N_5 device. The Si nanostructures were fabricated by a pseudo-Bosch deep reactive ion etching process, then tantalum oxide (Ta_2O_5) was coated onto Si by atomic layer deposition (ALD), and finally the Ta_2O_5 was nitrided in pure ammonia gas at 900 °C.

To gain a deeper understanding of charge transport within the thin Ta_3N_5 shells and to optimize the electrode

performance, we varied the thickness of the shell between 10 and 70 nm and tested for ferrocyanide oxidation as a facile photoelectrochemical reaction. We found that in a core-shell Si-Ta_3N_5 nanostructured device, photocurrent decreased with increasing Ta_3N_5 thickness, which indicates that charge extraction becomes problematic with thicker Ta_3N_5 films. To verify that decreasing photocurrent was due to poor charge transport in thicker films, we performed absorbed photon to current efficiency (APCE) measurements with Ta_3N_5 deposited on planar Si and quartz substrates. Again, we found that APCE decreased once the films were thicker than 30 nm, demonstrating that the minority carrier diffusion length is likely tens of nanometers.

Finally, to address the problem of Ta_3N_5 photodegradation and to demonstrate photoelectrochemical performance for water oxidation in basic electrolyte with the core-shell system, $CoTiO_x$ and NiO oxygen evolution catalysts were coated onto Si-Ta_3N_5 by dipcoating from a sol-gel and ALD, respectively. In terms of activity, we found that $CoTiO_x$ interfaces well with Ta_3N_5 and yields a high photocurrent density compared to NiO. With NiO we were able to achieve a photocurrent density close to that of $CoTiO_x$ after Fe incorporation into NiO (henceforth called Ni(Fe)O).[2] However, with $CoTiO_x$, the device is not stable; but with only ~1 nm of Ni(Fe)O we see a marked improvement in stability. The device loses 40-50% of its peak photocurrent over a 24 hour chronoamperometry test. Compared to the current Ta_3N_5 reports in the literature, the Ni(Fe)O catalyst affords noteworthy stability.[3-5] Therefore, in this work, we demonstrate a heterostructured Ta_3N_5 system with a tandem core-shell Si-Ta_3N_5 configuration that enabled earlier onset of photocurrent, efficient charge extraction from Ta_3N_5 and relatively stable long term water oxidation with a Ni(Fe)O co-catalyst.

1. Chun, W.-J.; Ishikawa, A.; Fujisawa, H.; Takata, T.; Kondo, J. N.; Hara, M.; Kawai, M.; Matsumoto, Y.; Domen, K. *The Journal of Physical Chemistry B* **2003,** 107, (8), 1798-1803.

2. Trotochaud, L.; Young, S. L.; Ranney, J. K.; Boettcher, S. W. *J Am Chem Soc* **2014,** 136, (18), 6744-6753.

3. Li, Y.; Zhang, L.; Torres-Pardo, A.; González-Calbet, J. M.; Ma, Y.; Oleynikov, P.; Terasaki, O.; Asahina, S.; Shima, M.; Cha, D.; Zhao, L.; Takanabe, K.; Kubota, J.; Domen, K. *Nat Commun* **2013,** 4.

4. Liao, M.; Feng, J.; Luo, W.; Wang, Z.; Zhang, J.; Li, Z.; Yu, T.; Zou, Z. *Advanced Functional Materials* **2012,** 22, (14), 3066-3074.

5. Liu, G.; Fu, P.; Zhou, L.; Yan, P.; Ding, C.; Shi, J.; Li, C. *Chemistry â€" A European Journal* **2015,** 21, (27), 9624-9628.

462505 Heterostructured c-Si/$BiVO_4$ Core-Shell Tandem Photoanode for Unassisted Photoelectrochemical Water Splitting

Abstract:

Photoelectrochemical (PEC) water splitting is a promising way of storing solar energy in hydrogen, which can be used as a carbon-free fuel or a chemical feed stock. In principle, a PEC electrode consists of a single active material capable of handling the key processes: absorption of solar photon energy, charge separation and transport, and electrocatalysis to make and break chemical bonds at the electrode surface. In practice, there is no existing material that can handle all of these functions simultaneously; thus, the multicomponent system that integrate different materials with more specialized functions is a more viable approach.

A tandem-junction PEC device offers a high theoretical solar-to-hydrogen (STH) efficiency due to the additive voltages across two photoabsorbers and the better utilization of the solar spectrum. Modeling STH efficiency as a function of band gap for a tandem device shows that over 20% STH can be achieved with band gaps of 1.2 and 1.8 eV.[1] One promising photoabsorber material combination is c-Si paired with a metal oxide, especially $BiVO_4$. In addition to the appropriate band structure alignment of both semiconductors, c-Si and $BiVO_4$ are low cost, earth abundant, and industrially scalable. However, the key challenges for these materials are the instability of c-Si in aqueous environment and the poor electronic properties of $BiVO_4$.

In this work, we present the fabrication of a heterostructured c-Si/$BiVO_4$ core-shell tandem based device capable of performing spontaneous water splitting without any precious metal. The wafer scaled device architecture was prepared by (1) nanostructuring of the c-Si substrate to serve as both a bottom absorber and a scaffold to improve the charge separation of $BiVO_4$; (2) tuning the c-Si p-n junction depth to maximize the device photovoltage; (3) engineering the SnO_2 interfacial layer to passivate the c-Si/$BiVO_4$ interface and protect the c-Si p-n junction; (4) conformally coating a $BiVO_4$ thin film on the nanostructured Si; and (5) decorating the c-Si and $BiVO_4$ surfaces with both oxygen and a hydrogen evolution catalysts.

(1) Seitz, L. C.; Chen, Z.; Forman, A. J.; Pinaud, B. A.; Benck, J. D.; Jaramillo, T. F. *ChemSusChem* **2014,** *7* (5), 1372-1385.

462617 The Fabrication of $Cu_2o/g\text{-}C_3N_4/WS_2$ Triple-Layered Photocathode for Photoelectrochemical Hydrogen Evolution

Abstract:

The Fabrication of $Cu_2O/g\text{-}C_3N_4/WS_2$ Triple-layered Photocathode for Photoelectrochemical Hydrogen Evolution

Xintian Xu, Yuanzhi Zhu, Xiaobin Fan, Guoliang Zhang, Wenchao Peng*

School of Chemical Engineering & Technology, Tianjin University, Tianjin 300072, China.

wenchao.peng@tju.edu.cn

Hydrogen is being vigorously pursued as a future energy carrier due to the fossil energy depletion. Photoelectrochemical (PEC) water splitting is an emerging hydrogen production technology by utilizing the solar light. This technology integrates in the same device both solar energy collection and water electrolysis. Both photo-active semiconductor and catalysts for hydrogen evolution reaction (HER) are necessary in this device. Cu_2O is a direct P-type semiconductor with a narrow band gap (~2.0 eV), which can be activated by the visible light. Moreover, Cu is a kind of earth abundant element, and the synthesis of Cu_2O is facile. Cu_2O is therefore one of the most promising low-cost photoactive semiconductor reported until now. However, the activity of bare Cu_2O is still poor, and it is not stable towards chemical oxidants and light corrosion. Combing with other semiconductors (e.g. TiO_2, ZnO) could improve the activity and stability of Cu_2O effectively.

As we know, Pt and other precious metals are the most efficient catalysts for hydrogen evolution reaction (HER).

However, these metals are rare and expensive to apply. Transition-metal dichalcogenides (TMDs), such as Molybdenum disulfide (MoS_2), tungsten disulphide (WS_2), are emerging HER catalysts that could be used as a substitute for noble metals in HER.

In this study, a novel $Cu_2O/g\text{-}C_3N_4/WS_2$ triple-layered photocathode was designed for the PEC hydrogen evolution. Cu_2O was electrodeposited on the FTO glass as the active layer for the light harvesting. A metal-free $g\text{-}C_3N_4$ was exfoliated and covered on the surface of Cu_2O as protection layer. $G\text{-}C_3N_4$ could also form p-n junction with Cu_2O to increase the activity. 1T-$WS_2$was obtained by lithium intercalation and used as HER catalyst for the further modification of the photocathode.

The obtained photocathode was then used for PEC HER test in the Na_2SO_4/H_2SO_4 solution (Î»â‰¥420nm, 100 mW/cm^2). With the irradiation, electron-hole pair will be generated in the p-type Cu_2O, and the excited electron could be then extracted through the $g\text{-}C_3N_4$ protective layer to produce hydrogen on the WS_2. The efficient extraction of excited electrons by this triple-layered photocathode leads to photocurrents of up to -6 mA/cm^{-2} at -0.4 V versus RHE (pH 6.0). The corresponding overpotential is -0.23 V (versus RHE) for the H_2 evolution. In sharp contrast, nearly no H_2 evolution happened under dark conditions. This photocathode is facile to synthesis and has great potential for developing low-cost PEC water-splitting device.

462862 1T-WS_2 on Graphite Foam As a Binder-Free Electrode for Enhanced Hydrogen Evolution

Abstract:

1T-WS_2 on Graphite Foam as a binder-free electrode for Enhanced Hydrogen Evolution

[1]Xiaomeng Guo, [1]Yuanzhi Zhu, [2]Junyi Ji*, [1]Xiaobin Fan, [1]Guoliang Zhang, [1]Fengbao Zhang, [1]Wenchao Peng*

[1]School of Chemical Engineering & Technology, Tianjin University, Tianjin 300072, China,

wenchao.peng@tju.edu.cn

[2]College of Chemical Engineering, Sichuan University, Chengdu, Sichuan 610065, China,

junyiji@scu.edu.cn

Hydrogen is regarded as a clean alternative energy carrier due to its high energy density and environmental friendliness. Electrochemical catalytic hydrogen evolution reaction (HER) is an efficient method for the production of hydrogen. Pt-group metals are the most effective HER catalysts due to their low overpotential and high efficiency. However, these metals are rare and expensive to apply on an industrial scale. To develop highly active and low-cost HER catalysts remains a big challenge. Molybdenum disulfide (MoS_2) and tungsten disulphide (WS_2) are emerging HER catalysts that could be used as a substitute for noble metals in HER.

In this study, a three dimensional graphite foam (GF) was synthesized on the Ni foam using chemical vapor deposition (CVD) method. After the removal of Ni foam with HCl aqueous solution, exfoliated 1T-WS_2 was dipped on the surface of GF to obtain the GF-WS_2 composite. Due to the 3D structure and strong mechanical strength of GF, the composite can be used as a binder-free electrode. The GF-WS_2 electrode was then tested for the HER performance in 0.5 M H_2SO_4 solution, and excellent HER activity was observed with a small overpotential of 0.1 V (vs RHE). Compared to the naked WS_2, the overpotential had been decreased ~0.1 V, and the cathodic current can increase ~10 times at a bias of -0.3V. This activity improvement should be due to the strong chemical and electronic coupling between the GF supports and WS_2 nanosheets. The electrical coupling could afford rapid electron transport from the less-conducting WS_2 nanosheets to the GF with more active sites. The GF-WS_2 is therefore a promising binder-free electrode with great potentials for hydrogen production.

462713 Flame Synthesis of Highly Transparent and Robust Nano-Layers for Enhanced Photo-Electrochemical Water Splitting

Abstract:

Direct chemical storage of solar energy is an attractive concept for off-the-grid renewable power generation. Currently the poor transparency of the thick low-cost catalyst layers required for water oxidation hinders the large-scale fabrication of efficient photo-electrochemical cells for artificial photosynthesis and hydrogen production. Flame synthesis of tailored nanoparticles and direct aerosol deposition is a scalable technology for the production of ultra-porous layers of photo-catalysts and optoelectronic devices[1]. Nevertheless the fragility[2] of this gas-phase self-assembly has limited their application in liquid environments with most studies requiring the in-situ high temperature annealing of the depositing particle to obtain sufficient mechanical stability [3]. This results in a drastic drop of the layer porosity and hinders the fabrication of optimal architecture for electrochemical systems.

Here, we report the engineering of robust high performance photo-electrodes with tunable porosity and composition. We discuss the critical parameters controlling the self-assembly of these nanoparticle layers and present a flexible approach for their mechanical and chemical stabilization. We apply this concept to the fabrication of photo- and electrochemical cells for water splitting demonstrating very high turn-over frequencies, controllable light absorption and efficient electron collection. We envision that this scalable synthesis approach can be readily implemented for the commercial production of low-cost devices for chemical energy storage and renewable fuel production.

References

1. Nasiri, N.; Bo, R.; Wang, F.; Fu, L.; Tricoli, A. *Adv. Mater.* **2015**,27, (29), 4336-4343.

2. Tricoli, A.; Graf, M.; Mayer, F.; Kühne, S.; Hierlemann, A.; Pratsinis, S. E. *Adv. Mater.* **2008**,20, (16), 3005-3010.

3. Tricoli, A.; Wallerand, A. S.; Righettoni, M. *J. Mater. Chem.* **2012,** 22, (28), 14254-14261.

468352 Investigation of the Use of Metal-Organic Frameworks for Combined Water Purification and Catalytic H_2 Production

Abstract:

Lately, a few studies have been reported on the use of metal-organic frameworks (MOFs) for water purification via adsorption of contaminants and/or via photocatalytic degradation. MOFs correspond to 3D crystalline nanomaterials networks consisting of metallic clusters connected via organic ligands. They are characterized by their high porosity and hence have been investigated for applications in water purification, especially for the removal of organics from water. It is interesting to notice that in parallel to those studies, recent work has also been devoted to using MOFs to photocatalytically produce H_2 from water and organic molecules. Up until now, organic contaminants adsorptive removal, photodegradation as well as H_2 evolution have been investigated independently of each other. Hence, an opportunity arises to combine water purification and energy generation (in the form of H_2).

In this work, we have focused on the investigation of two specific MOF families (UiO-66(Zr) and MIL-53) and investigated its adsorptive and photocatalytic properties. The UiO-66(Zr) material was synthesized along with its functionalized counterparts (X-UiO-66(Zr), with X = NH_2, NO_2, Br). MIL-53(M) was synthesized with different metals ions (M = Al, Fe and Cr). The materials were characterized by a number of analytical and spectroscopic techniques in order to determine the effects of functionalization on the band gap and conduction band. In addition, the materials were tested for the adsorption and photocatalytic degradation of organics in water and the simultaneous production of H_2. This work reports on the impact of the abovementioned structural and chemical modifications on the MOFâ€™s electronic properties and their performance as dual-purpose adsorbent/catalyst for water purification and H_2 production.

471808 $Ni(OH)_2$ As a Hole Mediator for Visible Light-Induced Urea Photo-Oxidation

Abstract:

Urea, a major component of human urine, has been proposed as a source of electrolytic hydrogen. We describe here CdS-sensitized TiO_2 assembled with urea electrocatalyst, $Ni(OH)_2$, on fluorine-doped tin oxide (FTO) coated glass ($FTO/TiO_2/CdS/Ni(OH)_2$) used as photoanode to realize simultaneous urea oxidation and water reduction using visible light irradiation. A photoelectrochemical reactor is employed to demonstrate the $Ni(OH)_2$ can serve as hole mediator to drive urea oxidation. The steady-state response of the semiconductor-catalyst electrode in the photoelectrochemical system is investigated, and the photoelectrode transient charge dynamics are studied by employing transient absorption spectroscopy. The obtained carrier dynamics can provide useful information on the competitive processes such as charge transfer and/or recombination within the photoanode.

470119 Study of Photocatalytic Activity of Nano Organic Hybrid Materials (NOHMs) for Photoelectrochemical Reduction of CO_2

Abstract:

A promising opportunity space for novel class of colloidal catalysts has emerged from recent advances in the synthesis of colloidal nanomaterials in which the composition of the inorganic core as well as the organic ligand shell can be precisely tuned. The prospect of tailoring the ligand shell, for optimized interaction with the surrounding fluid and the surface of the colloid, for catalytic conversion is particularly promising for photocatalytic reactions. We have investigated a new class of hybrid colloidal photocatalysts with combined functionality to capture CO_2 and convert it to liquid fuels. The photodriven CO_2-to-fuels reaction has profound technological implications as a prospect to harvest vast solar energy resources, capture CO_2 and store the energy in form of a liquid fuel. Scientifically, the photoreduction of CO_2 involves a complex interplay of coupled sub-processes including CO_2 transport to and adsorption on the catalyst surface, light absorption, proton-coupled multi-electron transfer, selectivity, and material stability. The photoelectrochemical reduction of CO_2 leads to various â€˜value-addedâ€™ products including methane, olefins, methanol, ethanol, formaldehyde, formic acid, formate, and carbon monoxide. Whereas access to novel colloidal nanoparticle catalysts has enabled advances in fundamental understanding of and control over their catalytic properties, progress towards deploying this know-how by integration with established CO_2 capture technologies (*i.e.*, amine scrubbing) has lagged behind. We embraced this challenge as an opportunity study basic processing-structure-performance relationships of nano organic hybrid materials (NOHMs) applied to the photoelectrochemical reduction of CO_2. We will describe the colloidal synthesis the inorganic nanoparticle core comprised of titania and metal co-catalyst and the CO2-philic ligand shell. We will discuss our recent advances in understanding and controlling the composition, morphology, ligand exchange, optical absorption, and CO_2 chemical and physisorption capacity. We will

relate structure and composition of the catalyst to the photocatalytic performance and will describe reaction products and conversion probed by gas and liquid chromatography.

472234 Nanostructured Silicon Photocathodes for Solar Water Splitting

Abstract:

Photoelectrochemical (PEC) water splitting is a promising route for synthesizing hydrogen using solar energy without direct CO_2 emissions.[1] PEC water splitting devices require highly active hydrogen evolution reaction (HER) catalysts to efficiently produce hydrogen at the photocathode.[2] Platinum and its alloys are the most active HER catalysts but the low availability of platinum may create problems for large scale implementation of solar water splitting.[3] There is a need to replace platinum with a non-precious metal earth-abundant alternative to enable cost effective, scalable devices.[4]

Recently, transition metal phosphides and sulfides have been shown to be among the most active nonprecious HER catalysts on a total electrode activity and turnover frequency (TOF) basis.[5] Though extremely active, the high surface area morphologies of these previously synthesized electrodes present additional challenges for integration onto PEC water splitting electrodes.[6] New methods are required to synthesize HER catalysts with a high density of active sites that also transmit a large portion of incident light.

We discuss our progress in synthesizing highly active transition metal phosphide and sulfide catalysts in nanostructured, low light-absorbing geometries. First, we compare nanostructured silicon semiconductors covered with optically thin catalysts to analogous flat silicon electrodes. These results provide insight into the general design considerations necessary to maximize the device efficiency for any given catalyst-absorber system. Next, we present a device structure which decouples the catalyst nanostructure from that of the silicon absorber. In so doing, we achieve low levels of recombination within the silicon and maximize the current-voltage characteristics of the device. Based on our findings, we discuss the application of our catalyst structure to other semiconductor systems and suggest general strategies for further improving the performance of solar water splitting electrodes.

[1] M. G. Walter, E. L. Warren, J. R. McKone, S. W. Boettcher, Q. X. Mi, E. A. Santori, N. S. Lewis, Chemical Reviews 2010, 110, 6446.

[2] L. C. Seitz, Z. Chen, A. J. Forman, B. A. Pinaud, J. D. Benck, T. F. Jaramillo, ChemSusChem 2014, 7, 1372.

[3] J. Greeley, I. Stephens, A. Bondarenko, T. P. Johansson, H. A. Hansen, T. Jaramillo, J. Rossmeisl, I. Chorkendorff, J. K. Nørskov, Nature Chemistry 2009, 1, 552.

[4] P. C. Vesborg, T. F. Jaramillo, RSC Advances 2012, 2, 7933.

[5] J. Kibsgaard, T. F. Jaramillo, Angew. Chem.-Int. Edit. 2014, 53, 14433.

[6] J. D. Benck, T. R. Hellstern, J. Kibsgaard, P. Chakthranont, T. F. Jaramillo, ACS Catalysis 2014, 4, 3957.

468999 Remarkable Enhancement of Photocatalytic Hydrogen Production in Sensitized and N_2/Ar Plasma Treated, Nanoporous TiO_2 Films

Abstract:

The optical and electronic properties of TiO_2 thin films provide tremendous opportunities in several applications including photocatalysis, photovoltaics and photoconductors for energy production. Despite many attractive features of TiO_2, critical challenges include the innate inability of TiO_2 to absorb visible light and the fast recombination of photoexcited charge carriers. In this study, we prepared ordered nanoporous TiO_2 films co-modified by graphene quantum dot sensitization and nitrogen doping (GQD-N-TiO_2) for hydrogen production from photoelectrochemical water splitting under visible light irradiation. First, cubic ordered nanoporous TiO_2 films were prepared by a surfactant templated sol-gel method. Then, TiO_2 films were treated with N_2/Ar plasma for the incorporation of substitutional N atoms into the lattice of TiO_2. GQDs were prepared by chemically oxidizing carbon nano-onions. The immobilization of GQDs was accomplished by reacting carboxyl groups of GQDs with amine groups of N-TiO_2 developed by the prior immobilization of (3-aminopropyl)triethoxysilane (APTES). Successful immobilization of GQDs onto N-TiO_2 was probed by UV-Vis, FT-IR, zeta potential and contact angle measurements. The nitrogen-doped TiO_2 (N-TiO_2) showed the significant reduction of band-gap (3.5 eV to 2.88 eV) and the unprecedented enhancement (240 times) of photocatalytic hydrogen production visible light illumination. The GQD-immobilized N-TiO_2 film showed about 2 times enhancement compared to N-TiO_2, showing the important role of GQDs. This outstanding enhancement is attributed to the high surface area of nanoporous films and synergistic effects of nitrogen doping and GQD sensitization resulting in enhanced visible light absorption, efficient charge separation and transport.

468774 Bifunctional Porous Materials for Combined CO_2 Capture and Catalytic Conversion

Abstract:

The rapidly-increasing concentration of Greenhouse Gases (GHGs) in the atmosphere has become a critical issue prompting the scientific and engineering community to find ways to efficiently mitigate emissions of GHGs, such as CO_2. Electricity generation using fossil fuels is one of the major contributors of CO_2 emissions and therefore, various end-of-the pipe technologies are currently proposed as carbon management media. A less energy intensive alternative than amine-based solvents are solid adsorbents which are capable of CO_2 capture at low temperatures as part of the Carbon Capture Utilisation and Storage process (CCUS). Another interesting development in CCUS research is the use of captured CO_2 as a chemical feedstock. Although this pathway will not eliminate the need for long term CO_2 storage, carbon utilisation could diversify the range of possibilities

Metal-organic frameworks (MOFs) represent a class of adsorbents with chemical and structural tunability, large porosity as well as high thermal and chemical stability. Owing to these features, MOFs are promising materials for CO_2 capture. In addition, recent research efforts point to the potential use of MOFs as photocatalysts, especially in the context of CO_2 reduction into useful products (e.g. CO and CH_4). Considering these two aspects, we are investigating the application of MOFs as bifunctional absorbent-photocatalyst materials for the combined capture and conversion of CO_2. This approach could potentially enable process intensification and improvement of the overall Carbon Capture Utilisation and Storage process sustainability. A series of MOFs were synthesised and fully characterised by a range of analytical and spectroscopic techniques (i.e. XRD, FT-IR, thermogravimetric analysis, N_2 sorption). These MOFs were tested as CO_2 adsorbents as well as heterogeneous CO_2 reduction photocatalysts in the gas phase using a purpose-built reactor. In the latter case, H_2 and H_2O as sacrificial agents under mild reaction conditions (ambient temperature and pressure). Products were analysed using a GC-MS and the MOFs were characterised after testing to identify any change in structure and/or chemistry. The results were used to identify mechanisms of reactions.

457454 Graphene Nanoribbons As Conductive Pathways in Directly Deposited Silicon Nanofiber Anodes for High Performance Lithium-Ion Batteries

Abstract:

Silicon is the best candidate for the next generation of Li-ion battery anodes due to its high specific capacity of 4200 mAh/g .However, it expands to 300% of its original size upon insertion of lithium ions[1], which results in crack formation and pulverization of active material with cycling and thus capacity fading. Crack formation also results in constant solid electrolyte interface formation on the surface of the newly exposed silicon, which brings about unstable performance. Several strategies such as use of nanoparticles[2] and composites[3] have been employed by research groups to overcome aforementioned issues. However, all the modifications to silicon has to be done in an economical manner for the product to become an industrial realization.

In the present work we are engineering the nanostructure of a silicon composite with graphene nanoribbons (GNRs) and polyvinyl alcohol (PVA) to create a reliable anode for lithium ion batteries. Silicon nanoparticles and GNRs are dispersed along PVA nanofibers via a water-based electrospinning process to directly form functional fiber mats on top of the copper current collector. This method is compatible with a role to role production, and bypasses several long-winded steps in the conventional paste-based electrode making. Physical and chemical properties of the nanoparticles and their dispersion affect the fiber morphology and ultimately electrochemical performance of the composite material. Graphene nanoribbons, made by unzipping carbon nanotubes, offer unique electrochemical and structural properties when inside the composite fibers. It was observed that GNRs outperform their precursor CNTs in terms of overall capacity and retention. The Silicon-GNR-PVA fiber anode exhibited an outstanding 2000 mAh/g at a rate of C/15 and 1200 at C/5 over 200 cycles. Inclusion of GNRs improved the structural stability and conductivity of the fibers, while protection of nanoparticles from the electrolyte offered by the polymer resulted in stable cycling of

the directly deposited electrodes.

[1] C.-M. Wang, X. Li, Z. Wang, W. Xu, J. Liu, F. Gao, L. Kovarik, J.-G. Zhang, J. Howe, D. J. Burton, Z. Liu, X. Xiao, S. Thevuthasan, D. R. Baer, *Nano Lett.* **2012**, *12*, 1624.

[2] M. M., Ryu I, Wu H, Liu N, Hu L, Nix WD, Cui Y Yao Y, *Nano Lett.* **2011**, *11*, 2949.

[3] Y.-S. Hu, R. Demir-Cakan, M.-M. Titirici, J.-O. Müller, R. Schlögl, M. Antonietti, J. Maier, *Angew. Chem. Int. Ed.* **2008**, *47*, 1645.

467753 Si-Based Nanofiber Anodes for Li-Ion Batteries Prepared Using Particle/Polymer Electrospinning

Abstract:

The demand for rechargeable batteries with high energy and power densities has never been greater. The functionality of many portable electronic devices is limited by battery size and lifetime. Similarly, electric vehicle propulsion systems require significant improvements to satisfy consumer expectations for long drive distances and short recharge times. To meet the demands of energy-hungry consumers, new battery electrodes with high capacities and long cycle life at fast charging rates must be developed.[1-2] These requirements have motivated research efforts to develop next-generation Li-ion battery materials, such as nano-structured silicon anodes. There is also a need for new electrode architectures to replace conventional slurry cast materials. One approach for an electrode design is the use electrospun electrodes containing engineered void spaces to enhance Li^+ transport rates for high areal capacities at fast charge/discharge rates.

Silicon has received considerable attention as an active material for Li-ion battery anodes due to its high theoretical capacity (3,579 mAh/g vs. 372 mAh/g for graphite). However, silicon based anodes typically suffer from rapid capacity fade due to particle cracking and electronic isolation caused by large volumetric changes between the lithiated and delithiated states. While particle fracture can be eliminated by using nanosized Si, electronic isolation remains problematic. Another significant obstacle for commercialization of Si in Li-ion batteries is the irreversible consumption of Li during formation of the solid-electrolyte interphase (SEI) layer.

The present work is the latest installment by Pintauro and co-workers on the use of particle/polymer electrospun nanofiber mats for fuel cell and Li-ion battery electrodes.[1-5] Such electrodes are a viable and attractive alternative to electrospun mats prepared by carbonizing polymer fiber precursors at high temperature. In particle/polymer electrospinning, there is no fiber pyrolysis; the electrospinning and post-processing are performed at room temperature which simplifies the process and preserves the beneficial characteristics of the polymer binder. Furthermore, the structure of the fibers can be easily altered for improved battery performance (e.g., by using core-shell or hollow-bore fibers or fibers with a high internal porosity). Due to these characteristics, high gravimetric, areal, and volumetric storage capacities can be achieved through the use of thick, densely packed particle/polymer nanofiber mats.

In this presentation, recent results will be presented on the fabrication and use of electrospun particle/polymer nanofiber anodes containing Si nanoparticles, carbon powder, and poly(acrylic acid) binder (Si/C/PAA). Charge/discharge properties of the nanofiber electrodes will be presented, and electrode capacities for different fiber mat thicknesses and mat porosities will be discussed. Notably Si/C/PAA nanofiber anodes have been prepared with high gravimetric, areal, and volumetric capacities of 1,100 mAh/g, 3.1 mAh/cm^2, and 560 mAh/cm^3, respectively, which were stable for 50 cycles at 0.1C. These results demonstrate that proper micron-scale design of electrodes can enhance Li-ion battery performance.

References

1. E. C. Self, R. Wycisk and P. N. Pintauro, *Journal of Power Sources*, **282**, 187 (2015).

2. E. C. Self, E. C. McRen, and P. N. Pintauro, *ChemSusChem*, **9**, 208 (2016).

3. W. Zhang and P. N. Pintauro, *ChemSusChem*, **4**, 1753 (2011).

4. M. Brodt, R. Wycisk and P. N. Pintauro, *Journal of The Electrochemical Society*, **160**, F744 (2013).

5. M. Brodt, T. Han, N. Dale, E. Niangar, R. Wycisk and P. Pintauro, *Journal of The Electrochemical Society*, **162**, F84 (2015).

463852 *In Situ* Activation of Nitrogen-Doped Graphene-Based Materials Anchored on Graphite Foam for High-Performance Energy Storage

Abstract:

Contact interface between the active materials and the current collector is essential for electron transfer rate and the mechanical properties of an electrode. With a traditional drop-casting electrode preparation method, the nano-sized active materials should mix with the nonactive and nonconductive binder, thus inevitably affect the total electrode resistance. Activation of graphene-based materials (e.g. graphene, carbon nanotube) by KOH can rebuild the crystalline structure to obtain continuous interface contact to facilitate electron transfer, while the molten KOH can act as the filler to enlarge the interlayer space to increase the electrolyte ion transfer rate.

Here, three-dimensional free-standing nitrogen-doped porous graphene-based materials/graphite foam composites (aNG/GF) were fabricated *via in situ* activation of nitrogen-doped graphene-based materials on highly conductive graphite foam (GF). During activation, the molten KOH reacts with the carbon to generate nano-holes as well as to rearrange the carbon, yielding a distribution of meso- and micro-pore channels and increases the specific surface area. After *in situ* activation, intimate 'sheet contact' was observed between the graphene sheets and the GF. The 'sheet contact' produced by *in situ* activation is found to be superior to the 'point contact' obtained by the traditional drop-casting method, and facilitates electron transfer. Due to the intimate contact as well as the abundant nano-holes and channels, the aNG/GF electrodes can exhibit high electron and electrolyte ion transfer rate, which is of great important for high-performance energy storage devices. The aNG/GF electrodes can deliver high gravimetric capacity and volumetric capacity with respect to the whole electrode mass and volume (including the active materials and the GF current collector). Furthermore, the tailored contact interface and the unique micro-structure could guarantee a high rate performance of the electrode even under ultrahigh current density, indicating its promising use as a high-rate LIB anode.

470624 Solution Combustion Synthesis for High Performance $ZnCo_2O_4$ Anode in Lithium-Ion Batteries

Abstract:

Nanostructured, tailored $ZnCo_2O_4$ is synthesized by solution combustion synthesis using two fuels, glycine and citric acid, to explore how reaction conditions impact electrochemical performance as a Li-ion anode. Owing to the unique combustion reaction mechanism, the novel strategy of using mixed fuels and ammonium nitrate creates anode material with improved cycling stability and capacity. Characterization shows a mesoporous spinel product, composed of nanoparticles assembled into micron-sized secondary structures, with properties varied by the fuel ratio and type. A stable capacity of 1,200 mAh g^{-1} is achieved at C/10 and 1000 mAh g^{-1} at 1C, showing high energy and power capabilities as compared to the bulk material. Based on an understanding of the reaction mechanisms and influence of reaction parameters, solution combustion synthesis is a powerful and flexible technique to produce superior electrode materials.

473388 a General and Mild Approach to Controllable Preparation of Manganese-Based Micro/Nanostructured Bars for High Performance Lithium-Ion Batteries

Abstract:

To meet the requirements for large-scale applications in electric vehicles, hybrid electric vehicles and stationary energy storage, growing interest has been directed at the economical, sustainable and environmental friendly methods to produce electrode materials for rechargeable lithium ion batteries (LIBs) with high energy and power densities.

Recently, manganese-based electrode materials, such as layered LiNixCo1-x-yMnyO2, spinel LiNi0.5Mn1.5O4 and layered Li-rich xLi2MnO3â€¢(1-x)LiNi1/3Mn1/3Co1/3O2 (0 â‰¤ x â‰¤ 1) have received a flurry of research activities due to their advantages of high energy and power densities, low cost and high cycling performance.

In respect to thermal stability, larger particles with a higher tap-density would be less reactive in a highly oxidized state. On the other hand, an electrode consisting of uniform particles with an appropriate size could ensure a homogeneous intercalation reversibility from particle to particle over repeated cycles. Therefore, the control of particle size and morphology for tailoring lithium-ion battery electrode materials has been paid more and more attention in recent years. Till now, it has remained a challenge to develop a general strategy that allows the large-scale preparation of a wide array of 1D micro/nanostructured manganese-based electrode materials with a controllable phase, composition and shape, as well as a narrow size distribution.

Herein we have established a Chimie Douce approach --- by mixed solvent co-precipitation --- to selectively prepare an extensive series of 1D micro/nanostructured manganese-based electrode materials, both cathode and anode for LIBs with high energy and power densities, attesting to its generality and effectiveness. Based on the kinetic elaboration of crystallite nucleation and growth through mixed solvent mediation, the new route is easily controllable, repeatable, and generalizable to the fabrication of a large series of some highly promising manganese-based electrode materials with tunable 1D micro/nanostructures for enhanced performances in lithium ion batteries. As a significant difference from the conventional methods, the present method is conducted in ethanol/water mixed solutions at room temperature, without having recourse to long-chain surfactants or hydrothermal condition. In addition, it can be employed to produce different aspect ratios of microbars with mixed transition metal oxides by tuning the ethanol/water volume ratio in the mixed solvent. It is anticipated that the mixed solvent mediated co-precipitation strategy will open opportunities to heuristically and systematically design mixed transition metal oxides with controllable phase and composition, including not only manganese-based but also other metal-based materials, with desirable micro-/nano-structures for developing high performance lithium-ion batteries.

460764 Advanced Energy Materials Derived from Bijels

Abstract:

Pressing energy demands for national security, distributed power, and portable technologies require the development of next-generation batteries and supercapacitors that can simultaneously deliver high power and energy densities. In order to achieve this goal, the kinetics of ion and electron transport within the electrodes must be enhanced while maintaining a large capacity for energy storage. Therefore, the idealized electrode microstructure has been envisioned as a three-dimensional, co-continuous arrangement of percolating passages for efficient ion and electron transport that would also enable a large volumetric density of active materials. We report a novel method to produce this distinctive microstructure through the use of soft matter templates derived from bicontinuous interfacially jammed emulsion gels (bijels), which are a new class of soft materials with a co-continuous microstructure where two interpenetrating fluid domains are kinetically arrested through jamming of particles at the fluid-fluid interface during spinodal decomposition. Bijels can be converted to electrolytically active composites with co-continuous internal microstructure through simple chemical post processing steps. In this talk, we will present microstructured composite electrodes that have been synthesized by this technique and demonstrate how their electrochemical characteristics can be tuned over a wide range, allowing for concurrent delivery of high energy and power. The broader applications of this new class of soft materials in electrochemical energy conversion and storage, sensing, catalysis, and tissue engineering will also be discussed.

457746 Sulfur Self-Doped Micro/Mesoporous Carbon Derived from Lignin and Its Application onÂ Supercapacitor Â and Oxygen Reduction Reaction

Abstract:

The growing global concerns about energy needs, fossil fuels consumption and the related environmental issues have motived scientists to find new, green and sustainable energy resources and technologies. In this work, byproduct lignin was successfully converted into self-sulfur-doped graphitic carbons via in situ hydrothermal carbonization (HTC) by following the post-annealing treatment at elevated temperature. The surface, porosity, textural, structural and morphological properties of synthesized materials were investigated by SEM, XPS, Raman, FTIR and surface area analyzer. Electrochemical properties of the prepared materials were analyzed by cyclic voltammetry (CV), charge-discharge (CD), electrochemical impedance spectroscopy (EIS), linear sweep voltammetry (LSE) in a KOH electrolyte solution. Up to 3.5 wt. % sulfur intercalating on carbon molecules was successfully achieved. The resulting materials electrode displays high surface area (up to 550 m^2 g^{-1}) with mesoporous structure and partially graphitic/amorphous carbon structure. The obtained original electrode materials result in high electrochemical activity for supercapacitor applications. The SGC-850 electrode exhibited a specific capacitance value of 225 g^{-1} at a scan rate of 1 A g^{-1}. The SGC-850 electrode revealed a very consistent response over 10000 cycles at a harsh condition of 5 A g^{-1} current density. In addition to supercapacitor application, the obtained material was tested for oxygen reduction reaction (ORR) application. The ORR CV curve of the SGC-850 presents well-defined cathodic peak in the O_2 electrolyte. The SGC-850 goes under 4 electron pathway for the O_2 reduction. To sum up, metal-free SGC-850 is promising candidate electrode materials for supercapacitors and ORR applications.

465633 Synthesis of Hollow Co_3O_4 Nanoparticles on Nitrogenâ€"Doped Porous Carbons for High-Performance Supercapacitors

Abstract:

Supercapacitors have been considered as next generation energy storage devices due to their high power density caused by rapid charge/discharge rates, long life cycle and eco-friendliness. In view of electrode materials, several strategies to improve performance of the supercapacitors were suggested and applied. We incorporated Co_3O_4 nanoparticles into nitrogen-doped porous carbons to investigate their potential as supercapacitors. The nano-sized Co_3O_4 particles grow and their morphology becomes hollow on the carbon support, which is attributed to the atomic diffusion (the Kirkendall effect) during the annealing. The cobalt-doped composites have high BET surface areas of about 3000 m^2 g^{-1} and possess mesopore-dominant structures caused by the hollow Co_3O_4 nanoparticles. As a result, the hollow Co_3O_4 in the composites simultaneously induces both enhanced pseudo-capacitance and efficient pore structure for energy storage. The specific capacitance of the composites is as high as 300 F g^{-1} in an aqueous solution of 6 M KOH at 1 A g^{-1}, and the material shows competitive specific capacitance of 210 F g^{-1} at 10 A g^{-1} for a high power density. The present study provides a unique insight into the synthesis of hollow nanoparticles on a porous substrate material for high-performance supercapacitance.

458006 Lecithin-Derived N-Doped Carbons for Supercapacitor Electrode Material

Abstract:

Soybean lecithin, plant derived surfactant, is a well-known natural surfactant in the animal tissue, food, and paints industry. Because of lecithin production is huge amount annually, it is urgent to find additional application of lecithin. In this work, an active supercapacitor electrode material is synthesized from lecithin-derived biomass material. Lecithin (as a carbon source) was first converted into bio-char via hydrothermal carbonization (HTC) then urea was introduced as a nitrogen source for biochar. Up to a 4 wt.% nitrogen intercalating into carbon was achieved. The morphology, thermal stability, and surface properties of lecithin, bio-char and N-doped carbon were examined. The resulting material show high surface area with micro/meso porosity. The capacitive characteristics of as-prepared materials were investigated by means of cyclic voltammetry (CV) (from 1.0 to 100 mV s^{-1}) and charge-discharge (from 1.0 to 20 A g^{-1} respectively. NGC-800 exhibited the highest capacitance value (411 F g^{-1} at 1 mV s^{-1} scan rate via cyclic voltammetry and 495 F/g at 1 A g^{-1} current density in 1 M KOH. Therefore, this study suggests that N-doped graphitic carbon materials synthesized from lecithin may be environmentally-sustainable candidate for energy storage applications.

452717 3D Porous Graphene Nanostructure Fabricated with a Simple, Fast, Scalable Process for Applications in High Performance Flexible Gel-Type Supercapacitors

Abstract:

A simple, fast, and scalable mix-and-heat process was developed for production of three-dimensional (3D) porous graphene nanostructure. The process involves only mixing and heating of starch and a graphene oxide (GO) suspension at 90 °C for 10 min to form 3D graphene monoliths, from which a three-dimensionally well-connected porous graphene nanostructure, starch/RGO, possessing a high specific surface area of 1519 m^2/g was obtained through common carbonization and activation treatments. No time-consuming hydrothermal or sol-gel chemistries and no energy-demanding supercritical fluid or freeze drying operations are needed to form the 3D graphene monoliths. The starch/RGO material was applied as the electrode material to fabricate flexible, gel-type symmetric supercapacitors of outstanding capacitive performances, delivering a high energy density of 19.8 Wh/kg at the power density of 0.5 kW/kg and exhibiting an excellent high rate capability of a high power density of 9.9 kW/kg at the energy density of 9.6 Wh/kg, among the highest for carbon material based, gel-type, symmetric supercapacitors. These excellent capacitive performances may be attributed to the advantageous structural features of the starch/RGO material, including the large surfaces exposed to the electrolyte ions for capacitance generation, highly porous structure for fast mass transfer of the electrolyte ions, and three-dimensionally well-connected conductive paths for the involved charge transport. The starch/RGO based supercapacitor exhibited outstanding mechanical stability with a retention rate of 90% in both energy and power densities at a large bending angle of 138°. It was also found that the starch/RGO based supercapacitor exhibited optimal performances in both energy and power densities at 80 °C and is thus applicable in a wide temperature range environment.

470638 Wet Spinning of Transition Metal Chalcogenide Fibers

Abstract:

Fibers comprising of molybdenum disulfide (MoS_2) were fabricated by a facile wet spinning procedure utilizing a coagulation bath of polyvinyl alcohol (PVA) for inclusion into flexible supercapacitors. Stable dispersions consisting of MoS_2, carbon nanotubes, and surfactant were prepared and characterized by Raman spectroscopy, transmission electron spectroscopy (TEM) and x-ray photospectroscopy (XPS). The latter of which indicates the presence of both exfoliated sheets and non-exfoliated nanoplatelets. Morphology of the fibers were characterized by scanning electron microscopy and electron diffraction spectroscopy (SEM/EDS) indicating that the exterior is principally comprised of MoS_2 nanoplatelets coated by carbon, the latter arising from the coagulation process. Meanwhile cross sections of the fiber reveal that the interior is predominantly MoS_2. The resultant fibers are durable, evident by the ability to be tied into knots, with mechanical properties similar to carbon nanotube fibers prepared in a similar fashion. Cyclic voltammetry measurements were performed in electrolyte solutions of H_2SO_4 and Na_2SO_4 in order to evaluate the ability of the wet spun fibers to serve as anodes in supercapacitor devices. Finally fibers prepared from tungsten disulfide (WS_2) were prepared to showcase the universality of this method towards other members in this class of transition metal chalcogenides (TMD).

468142 Dehydrogenation Mechanisms in Nanocrystalline Magnesium Hydride

Abstract:

Magnesium hydride is a promising material for on-board hydrogen storage due to its high gravimetric capacity (7.6 wt. %). However, its applicability is limited by slow dehydrogenation kinetics. Dehydrogenation mechanisms involve kinetics of several phenomena and the morphological changes of metal/hydride phase.

Dehydrogenation kinetics consists of nucleation and growth of magnesium phase by interfacial movement and/or by diffusion of H-atom through the hydride/Mg phase. Kinetics is studied using Johnson-Mehl-Avrami-Kolmogorov (JMAK) equation: α=1-exp(-kt^n), where α: converted fraction of Mg phase, k: kinetic factor involving nucleation density (N_0), growth velocity (U) and growth dimensionality (n) of Mg phase. n represents the morphological manifestation of Mg phase which can initially nucleate and grow as isolated colonies in 3D ($n\approx3$). Eventual impingement of the neighbouring Mg colonies, can arrest growth in the lateral dimensions (i.e. $n<3$). n and k along with other kinetic factors can explain the dehydrogenation mechanisms.

Dehydrogenation mechanism was investigated by performing isothermal experiments at 320-400 °C. The dehydrogenation curves show two distinct stages, namely, the incubation stage and post incubation stage. The incubation stage experiences slower kinetics and is observed predominantly at lower temperatures. Dehydrogenation mechanisms in the incubation and post incubation stages are studied by dividing the entire dehydrogenation curve into several small segments and applying JMAK analysis. This analysis yields discrete values of n and k which vary with time and their variation trends are different in the incubation and post incubation stages.

During the incubation stage, n increases from a negligible (n=0.01) to a significant (n=2.24) value. As n reflects the average growth dimensionality of the Mg phase, its increase can be attributed to the increase in number of growing Mg colonies with time. The activation energies for nucleation (10 kJ/mol) and growth (213 kJ/mol) are estimated from N_0 and U values, respectively. The activation energy of nucleation is one order lower than that for growth suggesting that Mg nucleation is fast while its growth is slow in the incubation stage.

In the post incubation stage, n values decrease with time from a high value (n>1) to a negligible value (n<0.50). The decreasing n can be attributed to the changing growth morphology of Mg phase during dehydrogenation. Microscopic observations (Dark Field TEM) and the estimation of volume fraction of Mg/MgH_2 phases shows that Mg phase nucleates and grows mainly near the particle surface and partly in the interior. During this stage, the interface velocity (U) changes by ~1 order. The estimated activation energies for nucleation and growth in post incubation stage are 55 kJ/mol and 152 kJ/mol respectively. This indicates that Mg phase growth limits dehydrogenation in the post induction stage.

The overall activation energy for dehydrogenation estimated using Raman spectroscopy is 158 kJ/mol. This is close to the activation energies for growth in both the incubation and post incubation stages, suggesting Mg phase growth as the controlling step during dehydrogenation of MgH_2.

471012 Heat Transfer in Metal Organic Frameworks during Gas Adsorption

Abstract:

Fast heat transfer during gas adsorption into metal-organic frameworks (MOFs) plays a key role in improving their efficiency for gas storage. Despite its importance, full understanding of heat transfer during gas adsorption into MOFs is still lacking. Using molecular modeling, we have investigated the transient phenomena related to heat and mass transfer within MOFs during gas adsorption. Our study reveals that heat transfer in MOF occurs much faster than gas diffusion. However, poor thermal transport within pure gas region prevents fast release of heat generated during adsorption. With respect to the effect of MOF size on its storage efficiency, we show that the density of adsorbed gas for bigger MOFs is lower due to higher temperature rises occurring during gas adsorption. Moreover, adsorption time is longer for bigger MOFs with respect to smaller MOFs.

458151 In-Vitro Dosimetry Model for Toxicity Ranking of Metal Oxide Nanoparticles

Abstract:

There is a debate in the nanosafety community regarding whether cellular toxicity of engineered nanoparticles (NPs) should be ranked on the basis of delivered dose (i.e., NP mass settled per suspension volume) or administered dose (i.e., initial NP mass concentration). Calculation of delivered dose with sufficient confidence for in-vitro toxicity ranking of NPs requires quantification of NP sedimentation considering fundamental suspension properties of NPs in cell culture media. Accordingly, we developed a first principles â€œparticles in a boxâ€ sedimentation model, which accounts for the particle size distribution, fractal dimension and permeability of agglomerated NPs. The developed sedimentation model was evaluated against both external and our own experimental sedimentation data for metal-oxide NPs, demonstrating significantly improved estimates of settled NP mass over other simplified sedimentation calculations with the assumption of impermeable and/or monodispersed NP agglomerates. The model was then applied to calculate the delivered NP doses in different cell culture media for the dose-response analyses of seven metal oxide NPs that were identified toxic from a library of twenty-four NPs in our previous studies. Hierarchical hazard ranking of the identified seven toxic NPs, according to the their EC_{50}, was not measurably different when determined on the basis of either delivered or administered doses. In contrast, simplified sedimentation calculations based on the assumption of monodispersed impermeable NP agglomerates led to significant underestimation of the settled NPs mass and thus distorted toxicity ranking.

472232 Amorphous Silicon Dioxide Nanoparticle Interactions with Pulmonary Epithelial Cells with and without a Pre-Existing Protein Corona

Abstract:

Silicon dioxide nanoparticles are increasingly abundant in consumer products and industrial applications, which increases their potential exposure to humans. In addition, these nanoparticles are under investigation for biomedical applications within the body such as targeted drug delivery. In either case, upon entering the body, nanoparticles interact with biological fluids and are immediately coated with proteins. Ultimately, it is this protein layer that interacts with the bodyâ€™s cells and tissues, and thus dictates the subsequent fate of the nanoparticles in the body. In this work, we investigated the adsorption of an abundant protein in biological fluids, albumin, to silicon dioxide nanoparticle surfaces and the impact of this protein layer on subsequent nanoparticle uptake within human cell cultures.

A previous mechanistic study conducted in our group has indicated that pH plays a strong role in the protein affinity for the nanoparticle surface (*publication in preparation*). Using attenuated total reflectance Fourier transform infrared (ATR-FTIR) spectroscopy, we obtained a real-time adsorption profile for BSA on the nanoparticle surface. Protein coverage on the nanoparticle surface was quantified using thermogravimetric analysis. Results from these studies indicated that at the isoelectric point of the BSA-silicon dioxide nanoparticle complex, pH 3.7, protein coverage was highest. Whereas at physiological pH, pH 7.4, much lower adsorption was observed over the same time period.

Cellular responses to uncoated particles and particles coated at these two pHs, to model a low-coating and a high-coating of protein on the nanoparticle surface, were conducted .Two pulmonary epithelial cell lines, 16HBE14o- and A549, were chosen for these studies as models of the different regions of the lung (bronchioles and alveoli) where nanoparticles deposit. Cell viability was measured via the MTT assay, particle uptake through TEM, and inflammatory markers through ELISA. Results from these studies indicated that coated particles behaved differently than uncoated particles in every parameter examined and that uptake depended on the cell system studied.

Overall, exposure to amorphous silicon dioxide nanoparticles induced a cellular response in pulmonary epithelial cells. Since these particles impacted the two cell lines differently, the deposition location of particle in the lungs after inhalation is important for understanding their toxicity. Surface functionalization prior to cellular exposure also played a key role in the cellular response, further supporting the idea that the protein corona affects cellular recognition and responses.

466807 Physiochemical Properties of Nanoparticles Determine Their *in Vitro* Cytotoxicity

Abstract:

Solid nanoparticles have proven useful in overcoming many of the barriers to drug delivery. Additional development has seen the attachment of a myriad of surface coatings such as polyethylene glycol (PEG). This commonly used PEG layer masks particles from innate host defence mechanisms within the body and increases circulation half-life of the therapeutic upon injection. Although this approach is widely accepted, the underlying mechanism is not well understood. This study examined the effect of hydrophilic surface coatings on solid silica nanoparticles with their interactions with Chinese Hamster Ovary (CHO) cells. Specifically, we examined the effect of differing lengths of PEG on intracellular uptake and viability.

In order to elucidate the effects of particle surface characteristics, studies were conducted with PEG 2k, 5k and 20k, as well as aminated silica nanoparticles. Two nanoparticle sizes, 60 and 120 nm, were chosen to draw size effect comparisons between the particles. These particles are generally recognized as a safe material by the Food and Drug Administration and provided a uniform foundation via size distribution, surface characteristics and mild reaction chemistry that precluded any nanoparticle-mediated effects.

In vitro distribution of the silica nanoparticles was found to directly correlate with their surface coating. Pegylated particles were found to remain interspersed among the CHO cells, whereas amine coated particles were found to produce punctate regions of fluorescence on and around the cells. Confocal microscopy confirmed uptake over simple surface association of the nanoparticles. Similar trends were correlated with flow cytometry. It was found that all PEG coatings decreased particle uptake by the CHO cells in comparison to those with an amine coating. The aminated particles had the greatest effect on cell growth. The 60 nm particles yielded greater half maximal inhibitory concentrations (IC_{50}) values over the 120 nm particles. The IC_{50} values for 60 nm aminated, PEG 2k, 5k and 20k particles were found to be 0.42 ± 0.18, 2.58 ± 0.09, 1.75 ± 0.11 and 3.535 ± 0.15 mg/mL, respectively. The IC_{50} values for the 120 nm aminated, PEG 2k, 5k and 20k particles were found to be 0.59 ± 0.05, 1.435 ± 0.13, 1.57 ± 0.125 and 1.333 ± 0.17 mg/mL. Growth inhibition directly correlated with particle concentration regardless of surface character. Pegylated 60 nm particles showed the greatest differences among their PEG coatings whereas the 120 nm pegylated particles showed similar results among all PEG sizes.

This study found that size and surface coatings have a dramatic influence on the cell viability and uptake *in vitro*. In addition, these studies show there may be an interplay between both of these variable sets. This work has provided a basis of comparison that could be extended to various nanoparticle systems. Understanding the impact of widely used solid nanoparticle compositions, sizes, coatings and their combinations on mammalian systems is an important step in developing robust drug delivery vehicles.

471757 The Impact of Titanium Dioxide Nanoparticles on the Lysosome-Autophagy System and Cellular Clearance

Abstract:

Titanium dioxide (TiO_2) is "generally regarded as safe" by the US Food and Drug administration and is found in a wide variety of consumer products, from cosmetics and pharmaceuticals to food additives. Recent studies suggest that the cellular uptake of nano-sized TiO_2 may cause oxidative stress, lysosomal impairment, and inflammation. However, the impact of nano-sized TiO_2 on cell physiology remains unclear. To address this knowledge gap, we evaluated the cellular response to the uptake of anatase TiO_2 nanoparticles (15, 50, and 100 nm in diameter), with particular emphasis on the lysosome-autophagy system, which is the cell's first response to the uptake of a material perceived as foreign and potentially harmful. Following characterization of the TiO_2 nanoparticles, cytotoxicity analyses were conducted and each of the three TiO_2 nanoparticles were found to induce minimal toxicity at concentrations in the cell culture medium up to 1000 µg/mL. Induction of the lysosome-autophagy system was then tested by monitoring the subcellular location of the transcription factor EB, a master regulator of lysosome- and autophagy-related genes. We found TiO_2 nanoparticles to induce activation of the transcription factor EB at concentrations as low as 10 µg/mL and to induce autophagic vesicle formation. However, uptake of high concentrations of TiO_2 nanoparticles (100 µg/mL and above) was also found to prevent vesicle turnover. We hypothesized that this observed effect on vesicle turnover was due to lysosomal impairment. Therefore, we analyzed lysosomal hydrolase function and membrane stability following TiO_2 nanoparticle uptake and observed a correlation between lysosomal function and integrity and vesicle turnover. In conclusion, we established that cellular uptake of TiO_2 nanoparticles induces activation of the lysosome-autophagy system, and thus upregulation of cellular clearance, but results in a blockage of autophagic flux and degradation *via* impairment of the lysosome.

458173 Potential Impact of Sublethal Levels of Nanomaterials on Interactive Behavior of Environmental Bacteria

Abstract:

Although bactericidal effects of a wide range of nanomaterials on environmental bacteria have been extensively reported, little is known about the impact of nanomaterials at sublethal levels on environmental bacteria and their interactive behaviors, which is a critical gap toward a comprehensive understanding of the impacts of nanomaterials on microbial ecosystems. In this study, using model environmental bacteria, we report the impact of single-wall carbon nanotubes (SWCNTs) and nanoSilver (nAg) on cell-to-cell interaction through influencing the production of â €œpublic goodsâ€ or the production of quorum sensing (QS) signals. Sublethal level of SWCNTs (40 or 80 Âµg/ml) inhibited the production of pyoverdine (PVD) (an important metabolite that involves in interactive behavior of microbial communities, especially in iron-limiting environments) in planktonic cultures of *Pseudomonas aeruginosa*. The inhibitory effect has been confirmed at the mRNA level by transcriptional assay and quantitative polymerase chain reaction (qPCR) analysis. We further explored the effect of SWCNTs on mature biofilms of *P. aeruginosa*. Our results showed that the exposure to SWCNTs for up to 48 h did not significantly influence PVD production in cells encased in mature biofilms; however, the exposure to SWCNTs promoted cell detachment from the biofilms and decreased PVD production of the detached cells. To explore the influence of nanomaterials on quorum sensing, we used *P. syringae* and *Pantoea stewartii* as model organisms and quantified the influence of SWCNTs and nAg on the production of different acylated homoserine lactones (AHL) signals. Our results showed for the first time that differential effects could be exhibited by nanomaterials on different QS signals. Our findings advance current knowledge toward a comprehensive understanding of the ecotoxicity of nanomaterials.

462954 Porous Structure Based Electrocatalysts for High Performance Fuel Cells

Abstract:

In this work, we systematically studied the effect of pore size and doping site position of metal-free catalysts in MEAs using well-defined ordered mesoporous carbon system to achieve highly active metal-free catalysts in MEA levels. At first, we synthesized mesosporous phosphorus doped ordered mesoporous carbon with large pore (~38 nm) by block copolymer assisted co-assembly. Then, addition nitrogen doping was conducted with 2 different conditions using different sized nitrogen precursors. Finally, we fabricated and tested MEAs that employ metal-free catalysts as a cathode catalyst. As a result, the MEA performance of synthesized catalyst with precisely controlled doping sites considerably exceeds that of previously reported metal-free catalysts.

465313 Design of Heterostructure Alloy Nanoparticles for Photocatalysis of CO_2 Reduction

Abstract:

Semiconductor nanocrystals of size in the strong quantum confinement regime, or quantum dots, exhibit size-dependent energy gap. The size-tunable shift of band-edge states can be a powerful tool in the control of exciton dynamics. For example, heterostructure nanocrystals with mere changes in geometric dimensions can lead to the band alignment from type I to type II, which could dramatically increase the exciton recombination lifetime.[1] In this presentation, I describe the design of heterostructure nanocrystals showing long exciton recombination lifetime. For the study, we synthesized various core/shell composite particles (e.g., PbSe/CdSe or CdSe/TiO2) and introduced CdS outer-most shell with varying thickness and morphology. [2] These heterostructure nanoparticles with prolonged exciton lifetime result in efficient photocatalysis of methylene blue reduction and hydrogen evolution from water.

In the way of addressing the heterostructures, we stumbled upon the observation that metal cocatalysts and surface can be engineered with relatively simple chemical approaches, rendering new photocatalytic systems suited in photocatalytic reduction of CO2 in gas phase. [3]

References

[1] Lee D C, Robel I, Pietryga J M and Klimov V I 2010 *J. Am. Chem. Soc.* **132** 9960

[2] Lee S, Lee K, Shin D J, Kim W D, Lee S and Lee D C 2014 *J. Phys. Chem. C* **118** 23612

[3] Jeong S, Lee S, Kim W D, Lee K, Lee S and Lee D C 2016 *ChemCatChem* in print

[4] Lee S, Jeong S, Kim W D, Lee K, Lee S and Lee D C 2016 *Nanoscale* in print

470971 Nanostructured Composite Intermediate-Temperature Solid Acid Fuel Cells Fabricated By Needleless Electrospinning

Abstract:

Fuel cells are the most efficient technology available today for the conversion of chemical energy into electricity. A new and promising fuel cell technology is the solid acid fuel cells (SAFCs), offering unique possibilities as a result of their intermediate-temperature region of fuel cells (100–300 °C). Their state of the art solid electrolyte, CsH_2PO_4 (CDP) is a solid acid compound with a superprotonic phase transition around 240 °C [1]. Under a dry condition the dehydration of CDP proceeds to form polymerization at temperatures above 200 °C, and CDP transforms into $CsPO_3$ [2]. Development of the catalytic performance can enable this type of fuel cell to be among the leading fuel cells on today's market.

Due to the pure ionic conducting nature of the electrolyte, the electrochemical reactions occur at the so-called triple phase boundaries, between the catalyst, electrolyte and the gas phase. Because of the triple phase, decreasing just the size of the catalyst does not increase the electrochemical reaction rates. However, in case the size of the electrolyte is also decreased, the electrochemical reaction rate is higher at the triple phase boundary resulting in increased power density of the fuel cell [1].

Needleless electrospinning technique was employed to fabricate nanostructured SAFC electrodes from $CDP/CsPO_3$-polymer solutions. Nanostructured CDP composite has the advantage of increased catalytic surface due to the large surface area. Needleless electrospinning is an inexpensive technique for producing nanostructured fuel cells at industrial scales. Instead of the commonly used nozzle, the needleless design employs a rotating drum immersed in a solution bath. Multiple Taylor cones form on the drum electrode surface resulting in rapid deposition of nanofibers over large areas. By applying $CsPO_3$ solutions, the phosphate groups form a chain, increasing the viscosity of the solution. This enables electrospinning with negligible amounts of polymers. The fibers produced from the solutions have diameters between 30 nm –

300 nm depending on the initial solution concentration, substrate temperature, and applied voltage. The deposition area is over 100 cm^2 and the deposition rate is up to 2 g/hr in the laboratory scale setup. Electrochemical characterization of the nano-composite electrodes, performed by A.C. impedance spectroscopy of symmetric cells under humidified hydrogen, confirmed the high activity of the electrospun nanocomposite electrodes. Fuel cell results using electrospun nanofibers from $CsPO_3$ solutions demonstrate that a 0.86 V cell voltage can be obtained at a current density of 50 mA/cm^2, which is ~10% higher than what can obtained using state-of-the-art techniques.

[1] C. R. Chisholm, D. A. Boysen, A. B. Papandrew, S. K. Zecevic, S. Cha, S., K. A. Sasaki, Á. Varga, K. P. Giapis, S. M. Haile, ***Interface***, 2009, ***18***, 53-59.

[2] Y. Taninouchi, T. Uda, Y. Awakura, A. Ikeda, S.M. Haile ***J. Mater. Chem.***, 2007, 17, 3182.

463690 Optimization of Pt/C Particles Electrocatalytic Activity By the Control of Carbon Nanostructures *Via* a Hybrid Aerosol-Colloidal Process

Abstract:

An optimum electrocatalytic performance of carbon-supported platinum (Pt/C) catalyst is determined by many factors, including the structure of the carbon support. In the study herein, we evaluated the effect of carbon nanostructures on the electrocatalytic performance of Pt/C catalyst and determined the most suitable nanostructure to deliver the best performance. Submicron carbon particles were prepared *via* spray pyrolysis of phenolic resin prior to being deposited with Pt nanoparticles using a microwave-irradiation process. Nanostructuration of the carbon particles was performed *in situ* owing to the colloidal interaction between phenolic resin and polystyrene latex (PSL) template, and was controlled by changing the properties of PSL. According to the results of the electrochemical analyses, Pt/C catalysts with porous carbon structure delivered the best performance compared to those with dense and hollow carbon structure. An evaluation on the pores size revealed the importance of a good balance between meso- and macropores for an effective electrocatalysis, which can be explained from the hydrodynamic profile in the catalyst layer.

469355 Highly Active Robust F Doped Transition Metal Oxide Based Solid Solution Electro-Catalyst for Acidic Medium Oxygen Evolution Reaction in PEM Based Water Electrolysis

Abstract:

The vast consumption of fossil fuels for meeting the energy demand has led to major unsustainable environmental concerns due to excessive greenhouse gas emissions warranting the identification and development of clean non-carbonaceous energy sources.[1-5] In this regard, hydrogen has been identified as a clean, non-carbonaceous and potential energy source with superior energy density than currently adopted carbonaceous energy sources for meeting the global energy demand. However, efficient and economic production along with cost-effective storage and distribution of hydrogen utilizing the clean non-carbonaceous approach is of paramount importance before universal adoption of hydrogen as a clean energy source in the impending non-carbonaceous fuel economy. Along these lines, electricity driven water splitting (water electrolysis) is a frontrunner among other clean pollution free approaches for hydrogen production (having low carbon footprint). The commercial development of water electrolysis is thwarted due to the need of expensive and precious noble metals based electro-catalysts (Pt, IrO_2) which exhibit excellent electro-catalytic activity and stability for oxygen evolution reaction (OER) in acid assisted proton exchange membrane (PEM) based water electrolysis. Hence, the identification and development of novel reduced noble metal containing electro-catalysts exhibiting excellent electro-catalytic activity and superior long term electrochemical stability similar/superior to state of the art OER electro-catalyst IrO_2 in harsh and highly acidic operating conditions of OER, will aid in the reduction of capital cost of water electrolysis cells and thus, its progression towards commercialization.[6]

With this aim in mind, exploiting theoretical first principles calculations of the total energies and electronic structures, fluorine doped transition metal (TM) oxide based solid solution electro-catalyst containing significantly lower noble metal content has been identified exhibiting surface electronic structure and electro-catalytic activity similar to that IrO_2.The thin films of 10 wt. % F doped $(TM_{1-x}Z_x)O_2$ (x=0.2, 0.3, 0.4; Z is a noble metal) coated on Ti foil were synthesized and studied as electro-catalyst for OER in acidic media. **Fig. 1** shows the SEM micrograph showing the nanoscale architecture of the representative composition $(TM_{0.8}Z_{0.2})O_2$:10F. Elemental x-ray maps show homogeneous distribution of elements throughout the material without any segregation at specific site.

Electrochemical characterization of the synthesized electro-catalysts has been carried out in a three-electrode configuration using 0.5 M sulfuric acid (H_2SO_4) solution as a proton source as well as the electrolyte, Pt wire as counter electrode and Hg/Hg_2SO_4 as the reference electrode (+0.65 V with respect to normal hydrogen electrode, NHE), with a scan rate of 10 mV/sec and at temperature of 40^0C. $(TM_{1-x}Z_x)O_2$:10F (for x = 0.2, 0.3, 0.4) electro-catalysts (total loading = 0.3 mg/cm^2) exhibit excellent electrochemical activity for OER with significantly reduced onset potential (~1.35 *vs* RHE) compared to that of IrO_2 (~1.43 *vs* RHE) (total loading=0.3 mg/cm^2). In addition, $(TM_{1-x}Z_x)O_2$:10F displayed outstanding ~8, ~14 and ~15 fold improved electro-catalytic activity at ~1.45 V (*vs* RHE) compared to IrO_2. Following chronoamperometry test conducted in 0.5 M H_2SO_4 solution at ~1.55 V (*vs* RHE) for 24 h, minimal loss in current density as well as excellent long term electrochemical stability was observed for $(TM_{0.8}Z_{0.2})O_2$:10F, similar to that of IrO_2.

These results show the promise of $(TM_{1-x}Z_x)O_2$:10F electro-catalyst portending ~80% reduction in noble metal content displaying significantly much higher electro-catalytic activity than IrO_2 and superior long term stability. In addition, these results show the promise of the formation F-doped solid solution in improving the electronic, physical, chemical and electro-catalytic properties of the doped TM oxide structures in acidic media. This will aid in significant reduction in the overall capital cost of PEM based water electrolysis for efficient and economic production of hydrogen. The results of these studies will be presented and discussed.

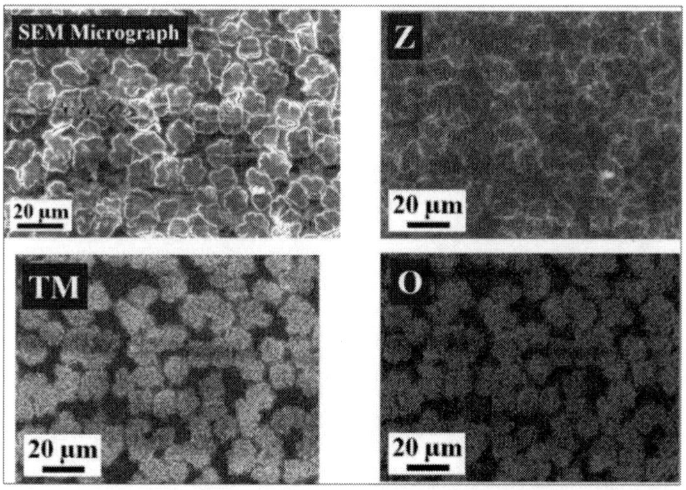

SEM micrograph with elemental x-ray maps of $(TM_{0.8}Z_{0.2})O_2$:10F electro-catalyst

References:

1 Patel, P. P. *et al.* Nanostructured robust cobalt metal alloy based anode electro-catalysts exhibiting remarkably high performance and durability for proton exchange membrane fuel cells. *Journal of Materials Chemistry A* **3**, 14015-14032 (2015).

2 Patel, P. P. *et al.* WO_3 based solid solution oxide–promising proton exchange membrane fuel cell anode electro-catalyst. *Journal of Materials Chemistry A* **3**, 18296-18309 (2015).

3 Patel, P. P. *et al.* High performance and durable nanostructured TiN supported Pt 50–Ru 50 anode catalyst

for direct methanol fuel cell (DMFC). *Journal of Power Sources* **293**, 437-446 (2015).

4 Patel, P. P. *et al.* Vertically aligned nitrogen doped $(Sn,Nb)O_2$ nanotubes – Robust photoanodes for hydrogen generation by photoelectrochemical water splitting. *Materials Science and Engineering: B* **208**, 1-14, (2016).

5 Patel, P. P. *et al.* Nitrogen and cobalt co-doped zinc oxide nanowires – Viable photoanodes for hydrogen generation via photoelectrochemical water splitting. *Journal of Power Sources* **299**, 11-24, (2015).

6 Datta, M. K. *et al.* High performance robust F-doped tin oxide based oxygen evolution electro-catalysts for PEM based water electrolysis. *Journal of Materials Chemistry A* **1**, 4026-4037, (2013).

Acknowledgements:

The authors gratefully acknowledge the financial support of NSF-CBET grant# 1511390. The authors also acknowledge the Edward R. Weidlein Chair Professorship funds and the Center for Complex Engineered Multifunctional Materials (CCEMM) for partial support of this research.

467668 Superwetting Nanoarray Electrodes for Gas-Involved Electrocatalysis

Abstract:

Superwetting Nanoarray Electrodes for Gas-Involved Electrocatalysis

Yingjie Li, Wenwen Xu, Zhiyi Lu, Xiaoming Sun*

Beijing University of Chemical Technology, Beijing 100029, China, Email: sunxm@mail.buct.edu.cn

Electrochemical gas-involved reactions can be divided into two types: gas-evolution reaction and gas-consumption reaction, both of which are crucial for a variety of energy conversion processes and industries (e.g. HER and ORR). For gas-evolution reaciton, if generated gas bubbles pin at the electrode surface and cannot escape from the surface in time, the accumulated bubbles will reduce the effective electrode surface area, increase diffusion resistance and enlarge polarization effect, resulting in more energy consumption. How to construct a novel electrode to promote gas bubble release is critical for improving the electrochemical efficiency besides activity improvement. Inspired from bio-inspired superwetting surfaces, we found that the interface behavior of electrode could be tuned by surface architecture construction, for example, transferring from aerophobic to superaerophobic by engineering a series of nanoarray electrode, eg. MoS_2 (Fig. 1), pine-shaped Pt, NiFe LDHs and Cu films [1-7]. This kind of superaerophobic electrodes could decrease the critical size of gas overflowing from the surface by cutting the three phase contact lines into discontinue dots, and thus reduce the diffused impedance and maintain the integrity of the solid−liquid interface that is necessary for electrocatalysis (e.g. water splitting and hydrazine fuel cells). On the other hand, for gas-consumption reaction, the construction of nanoarray ¡°superaerophilic¡± electrode could accelerate gas diffusion to reaction zone via gas-phase to solve the issue of low solubility and slow diffusion of gases in traditional electrocatalysiss reaction system with limited current density (eg., ORR). Therefore, the construction of superwetting nanoarray electrodes is imperative to improve gas transport at electrodes surface and to enhance activity and stability of electrodes.

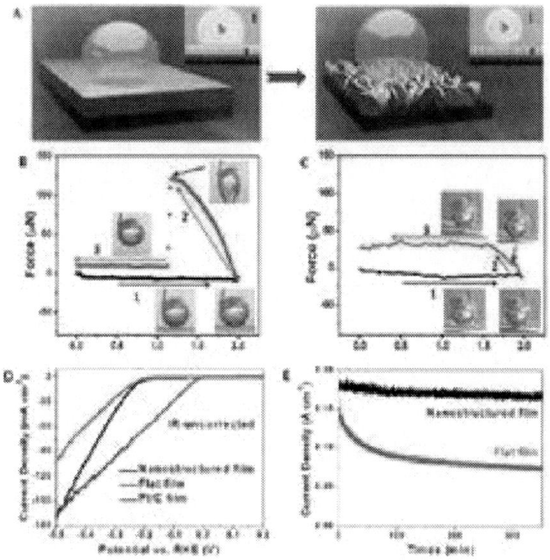

Fig. 1 (A) Schematic illustration of adhesion behaviors of gas bubbles on flat film (left) and nanostructured film (right), (B) and (C) Adhesive forces measurements of the gas bubbles on flat and nanostructured MoS_2 films, (D) Polarization curves (without IR-correction) of MoS_2 and Pt/C catalysts (E) Stability testing of flat and nanostructured MoS_2 electrodes

References

[1] Z. Lu, W. Zhu, X. Yu, H. Zhang, Y. Li, X. Sun, X. Wang, H. Wang, J. Wang, J. Luo, X. Lei, L. Jiang. Adv. Mater. 26 (2014), 2683-2687.

[2] Y. Li, H. Zhang, T. Xu, Z. Lu, X. Wu, P. Wan, X. Sun, L. Jiang. Adv. Funct. Mater. 25 (2015), 1737-1744.

[3] Z. Lu, M. Sun, T. Xu, Y. Li, W. Xu, Z. Chang, Y. Ding, X. Sun, L. Jiang. Adv. Mater. 27 (2015), 2361-2366.

[4] X. Liu, Z. Chang, L. Luo, T. Xu, X. Lei, J. Liu, X. Sun. Chem. Mater., 26 (2014), 1889-1895.

[5] M. Sun , Z. Lu , L. Luo, Z. Chang , X. Sun. Nanoscale, 8 (2016), 1479-1484

[6] W. Xu, Z. Lu, P. Wan, Y. Kuang, X. Sun. Small, 12 (2016), 2492¨C2498

[7] M. Jiang, Y. Li, Z. Lu, X. Sun, X. Duan. Inorg. Chem. Front. 2016, DOI: 10.1039/C5QI00232J.

Biography

Professor Xiaoming Sun gained his B.S. degree and Ph.D. in Department of Chemistry, Tsinghua University in 2000 and 2005, respectively. After postdoctoral work at Stanford University, he joined State Key Laboratory of Chemical Resource Engineering, Beijing University of Chemical Technology at 2008. His main research interests focus on separation and assembly of inorganic nanostructures, synthesis and separation of carbon nanomaterials and their composites, and structure control, opto-/electro-property investigations of nanoarrays and superwetting nanoarray electrodes for energy sciences. He has authored 63 journal articles (eg. ***J. Am. Chem. Soc., Angew. Chem. Int. Ed., Adv. Mater., ACS Nano.***), which have been cited >4600 times.

460470 Nanocatalysts for Water Splitting

Abstract:

Developing a clean and sustainable energy system is essential for the earth's sustainability. The hydrogen/water energy cycle, a clean and sustainable system, receives of great interest recently. Water splitting is one of the key reactions in the hydrogen/water energy cycle. Developing active and cost-effective nanocatalysts for water splitting is important to water splitting for the production of clean H_2 energy sources. This work will present various nanocatalysts for solar-driven and electric-driven water splitting reactions, especially for hydrogen evolution reaction (HER). The co-catalysts decorated Cu_2O/CuO photocathodes and MO/M heterostructured electrocatalysts were prepared by various interfacial engineering techniques. Combining advanced characterization and DFT computational techniques is capable of providing the insight of the heterostructured nanocatalysts for HER. It was found that the heterostructured interface plays a vital role and provides a highly active catalyst for HER. The design strategy for the HER catalysts can be developed via the understanding of the interfacial properties and reactions.

470723 Hydrogen Generation from Hydrous Hydrazine over Nickel-Doped Ceria Catalysts Prepared By Solution Combustion Synthesis

Abstract:

Hydrous hydrazine is considered to be a promising hydrogen storage material due to its high hydrogen content (8.0 wt.% for hydrazine monohydrate) and simple byproduct (only nitrogen) via complete decomposition. The key to exploit hydrous hydrazine in practical hydrogen generation applications is to develop efficient, selective, and economic catalysts for the decomposition. In this work, nickel-doped cerium oxides as new non-noble metal catalysts were prepared by solution combustion synthesis (SCS) of CeO_2, followed by impregnation of Ni or by one-step SCS of Ni/CeO_2. The effect of catalyst preparation conditions, including metal weight percentage (10%-80%), type of fuels (glycine, hydrazine, ethylene glycol, etc), ratio of fuel to oxidizer (0.5-3), and some additives on the properties of catalysts were investigated to optimize synthesis conditions and then their catalytic performance were tested. The experimental results demonstrate that the Ni/CeO_2 catalysts via SCS possess high catalytic activity and selectivity to generate hydrogen from hydrous hydrazine.

472204 Halide Perovskite Nanoplatelets

Abstract:

Colloidal perovskite nanoplatelets are a promising new class of semiconductor nanomaterials, exhibiting bright luminescence, tunable and spectrally narrow absorption and emission features, strongly confined excitonic states, and facile colloidal synthesis. Here, we demonstrate the high degree of spectral tunability achievable through variation of the cation, metal, and halide composition as well as nanoplatelet thickness. We synthesize nanoplatelets of the form $L_2[ABX_3]_{n-1}BX_4$, where L is an organic ligand (octylammonium, butylammonium), A is a monovalent metal or organic molecular cation (cesium, methylammonium, formamidinium), B is a divalent metal cation (lead, tin), X is a halide anion (chloride, bromide, iodide), and n-1 is the number of unit cells in thickness. We show that variation of n, B, and X leads to large changes in the absorption and emission energy, while variation of the A cation leads to only subtle changes but can significantly impact the nanoplatelet stability and photoluminescence quantum yield. Furthermore, mixed halide nanoplatelets exhibit continuous spectral tunability over a 1.5 eV spectral range, from 2.2 eV to 3.7 eV. These results demonstrate the versatility of colloidal perovskite nanoplatelets as a material platform, with tunability extending from the deep UV, across the visible, into the near-IR. In particular, the tin-containing nanoplatelets represent a significant addition to the small but increasingly important family of lead- and cadmium-free colloidal semiconductors.

467326 Ultrafast Photoexcited Carrier Dynamics in Ligand-Exchanged PbSe Nanocrystal Films: Lifetime, Mobility, Diffusion, and Interfacial Charge Transfer

Abstract:

Colloidal semiconductor nanocrystals have been used as building blocks for electronic and optoelectronic devices ranging from field effect transistors to solar cells. Properties of the nanocrystal films depend sensitively on the choice of capping ligand to replace the insulating synthesis ligands. Thus far, ligands leading to the best performance in transistors result in poor solar cell performance, and vice versa. To gain insight into the nature of this dichotomy, we used time-resolved terahertz spectroscopy measurements to study the mobility and lifetime of films of monodisperse lead selenide (PbSe) nanocrystals prepared with five common ligand-exchange reagents. Non-contact terahertz spectroscopy measurements of conductivity were corroborated by contacted van der Pauw measurements of the same samples. The films treated with different displacing ligands show more than an order of magnitude difference in the peak conductivities and a bifurcation of time-dynamics. Inorganic chalcogenide ligand-exchanges with sodium sulfide (Na_2S) or ammonium thiocyanate (NH_4SCN) show high THz mobilities above 25 $cm^2V^{-1}s^{-1}$, which is desirable for transistors, but nearly complete decay of transient photocurrent within 1.4 ns. The high mobility with NH_4SCN and Na_2S exchanges is more than offset by their short lifetimes and results in diffusion lengths of only ~200 nm. In contrast, ligand exchanges with 1,2-ethylenediamine (EDA), 1,2-ethanedithiol (EDT), and tetrabutylammonium iodide (TBAI) show ~5x lower mobilities but much longer carrier lifetimes, with ~30% of photoexcited carriers remaining for >10 ns. The long lifetimes with EDA, EDT, and TBAI yield diffusion lengths of at least 500 nm, which is approaching the film thickness desired for strong light absorption in solar cells. This bifurcated behavior may explain the divergent performance of field-effect transistors and photovoltaics constructed from nanocrystal building blocks with different ligand exchanges. Moreover, we have investigated ligand-exchanged PbSe films comprised of two different

nanocrystal sizes. We will report on the effects of band alignment, ligand chemistry, and structural order on interfacial charge transfer.

Reference: Guglietta, Diroll et al., *ACS Nano*, 2015.

470983 Band Edge Engineering of Hydroxide Nanoparticles for Semiconductor and Electrochemical Applications

Abstract:

Band Edge Engineering of Hydroxide Nanoparticles for Semiconductor and Electrochemical Applications

Matthias J. Young, Nicholas M. Bedford, Tatyana Kiryutina, Taylor J. Woehl

The band edge properties of a material largely determine its viability as an active material for optoelectronic, photochemical, and electrochemical applications, among others. Layered double hydroxides (LDHs) are a class of semiconducting materials which have garnered recent interest due to their (photo)catalytic, electrochemical, and corrosion-resistant properties. In this work, we examine the effect of cation composition on the band gap and band edge positions in a range of LDH structures. We employ atomistic materials simulations to model the band edge properties of LDH structures with structures comprised of various +2 cations (*e.g.* Co,Mg,Ni) and +3 cations (*e.g.* Al,Cr,V).This modelling is compared and contrasted with the experimental optical properties of LDH nanoparticles. We also explore the effect of substitutional dopants and point defects on the band edge properties of Mg-Al LDHs. LDH nanoparticles are synthesized using coprecipitation and their structural properties are evaluated using TEM and synchrotron characterization techniques. A subset of these LDH structures are also evaluated for their electrochemical and optoelectronic properties.

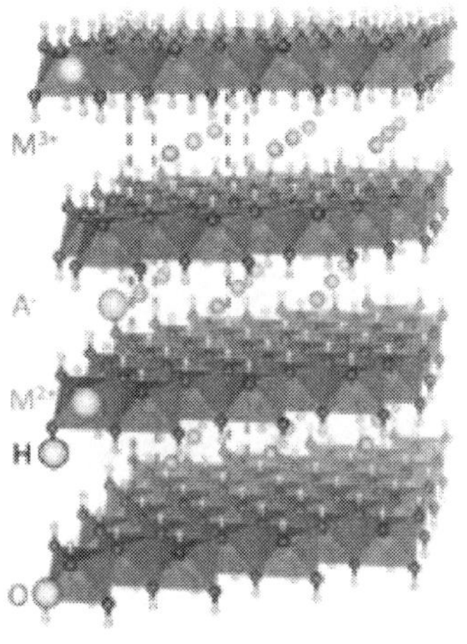

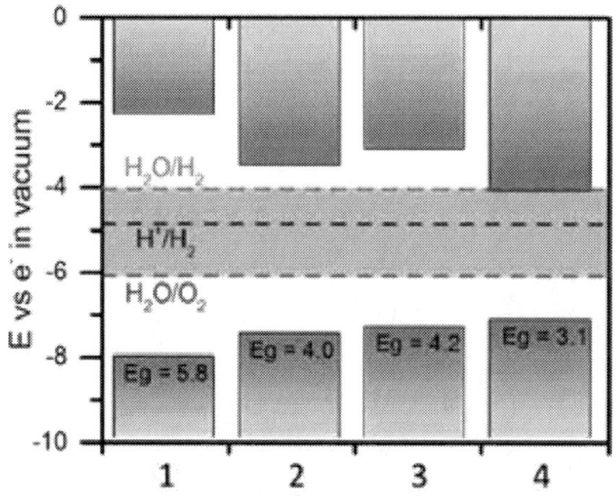

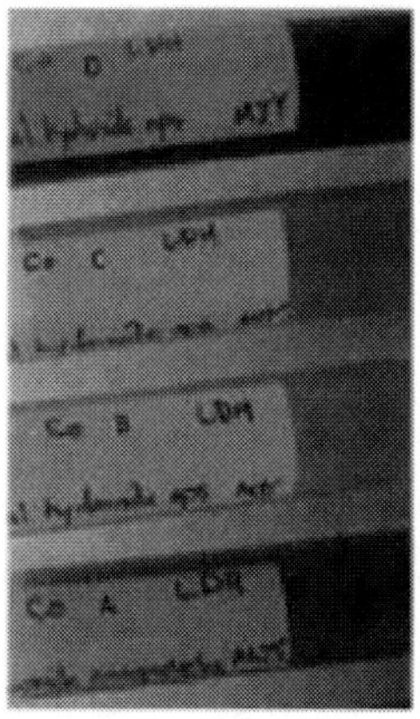

468498 Synthesis and Characterization of Cu_3SbS_4 Nanoparticles for Solution-Based Thin Film Solar Cells

Abstract:

The most efficient inorganic, thin film-based solar cells use CdTe or $CuInGaSe_2$ (CIGS) as absorber layers, both of which rely on toxic (Cd) and/or scarce elements (In, Te). The search for more sustainable processing of solar technologies has led to the identification of many Earth abundant and non-hazardous copper chalcogenide absorber materials. The Cu-Sb-S (CAS) system is promising in that it can be obtained in four major phases including $CuSbS_2$ (chalcostibite), $Cu_{12}Sb_4S_{13}$ (tetrahedrite), Cu_3SbS_3 (skinnerite), and Cu_3SbS_4 (famatinite), and the bandgaps for these compositions can vary from 0.5 to 2.0 eV.[1] The band-gap range and the high absorption coefficients (~ 10^5 cm^{-1} for Cu_3SbS_4 and 4x10^5 cm^{-1} for bulk $CuSbS_2$) make these materials of interest as a p-type absorber for thin-film solar cells.[2]

There have been several attempts to prepare CAS thin films to fabricate single junction solar cells, where fairly low efficiencies between 0.46 % for sputtered thin film Cu_3SbS_4 solar cells and 3.22 % for $CuSbS_2$ have been obtained.[3,4] One of the challenges for the Cu-Sb-S system is the poor thermal stability of CAS particles. Decomposition of the most promising phases, such as Cu_3SbS_4 and $CuSbS_2$, during thermal treatments can significantly change the films electrical and optical properties, which then limits the solar cell efficiency.[5]There are still many strategies that need to be investigated for the CAS absorber layers to reduce film defects and film decomposition, which should increase efficiencies for CAS-based solar cell devices.

Typically higher efficiencies are obtained for photovoltaic devices prepared using vacuum deposition techniques. This solid state approach can be difficult since the fine control over the final film composition is challenging for CAS materials, since many phases can coexist due to their nearly identical thermodynamic stability.[1] However, for CAS and Cu_2ZnSnS_4(CZTS) the highest efficiencies have been obtained for solution processed devices. For these systems, the use of nanoparticle inks to prepare thin film based solar cells may be a cost-effective approach for producing higher performance solar cells.

There are many challenges related to the solution-based synthesis of uniform CAS nanoparticles with controlled properties. We have used the hot-injection method to produce CAS, since the method provides a high level of supersaturation of reagents for a very short instant of time causing rapid formation of nuclei.[6] This nucleation step and the use of proper capping agents lead to the formation of monodisperse and uniform nanoparticles. In this study, we have synthesized colloidal famatinite (Cu_3SbS_4) nanoparticles. X-ray diffraction results indicated that optimization of the synthesis conditions allows selective formation of the Cu_3SbS_4 phase without other impurities. Thin films were deposited by spin-coating nanoparticle inks, and their optical and electrical properties were studied. Optimization of post-deposition annealing conditions was performed to obtain large grain films of the desired phase. Characterization of the CAS nanoparticles and films was performed using scanning electron microscopy, transmission electron microscopy, Raman spectroscopy, and UV-Vis-NIR spectroscopy to evaluate effects of synthesis conditions on the nanoparticle properties and also to evaluate the optimization of films obtained by post-deposition annealing steps. The goal of these studies is to determine the most effective method to enhance CAS film optical and electrical properties to achieve high efficiency CAS thin film solar cells.

References

1. Ramasamy, K., *et al.*, "Selective Nanocrystal Synthesis and Calculated Electronic Structure of All Four Phases of Copper–Antimony–Sulfide," *Chemistry of Materials,* **26,** pp. 2891–2899 (2014).

2. Van Embden, J. and Tachibana, Y., "Synthesis and characterisation of famatinite copper antimony sulfide nanocrystals," *Journal of Materials Chemistry,* **22,** pp. 11466-11469 (2012).

3. Franzer, N. D., *et al.*, "Study of RF sputtered Cu_3SbS_4 thin-film solar cells," *Photovoltaic Specialist Conference (PVSC),* 2014 IEEE 40th, pp. 2326–2328 (2014).

4. Banu, S., *et al.*, "Fabrication and characterization of cost-efficient $CuSbS_2$ thin film solar cells using hybrid inks," *Solar Energy Materials and Solar Cells,* **151,** pp. 14–23 (2016).

5. Welch, A. W., *et al.*, "Self-regulated growth and tunable properties of $CuSbS_2$ solar absorbers," *Solar Energy Materials and Solar Cells,* **132,** pp. 499–506 (2015).

6. Kwon, S. G. and Hyeon, T., "Formation Mechanisms of Uniform Nanocrystals via Hot-Injection and Heat-Up Methods," *Small,* **7,** pp. 2685–2702 (2011).

473898 Mesoscale Modeling of Stress-Directed Compositional Patterning in Semiconductor Alloys

Abstract:

The ability to scalably fabricate periodic, large-area assemblies of Ge quantum dots (QDs) on Si or SiGe substrates with high degree of spatial and size uniformity would impact numerous technologies, including nano-/micro-electronics, optoelectronics, nanosensor arrays, and high-density patterned media for data storage. In this talk, we study computationally a process in which SiGe substrates are compositionally patterned over large areas using spatially-modulated elastic fields applied by a nano-indenter array [1]. In this process, an indenter array is pressed against a SiGe wafer and the entire assembly is annealed at high temperatures during which the larger Ge atoms are selectively driven away from areas of compressive stress. Compositional analysis of the substrates demonstrates that this approach leads to a transfer of the indenter array pattern into the near-surface compositional distribution.

We describe here a mesoscopic model of the â€˜stress transferâ€™ process. The model is based on a multiresolution lattice kinetic Monte Carlo (LKMC) simulation that is propagated using rates for atomic diffusion that depend explicitly on local values of stress, composition, and temperature [2]. The dependence of atomic diffusion on composition is regressed to experimental data while the stress dependence is described using the theory of activation volumes [3]. The stress field is computed using a continuum linear elastic description of the contact problem and is updated quasi-statically as the composition in the SiGe substrate evolves.

The model is used to investigate systematically the impact of several process parameters, including the shape of the indenters, the spacing between indenters, the annealing temperature, and the indenter strength. We find that certain process parameter configurations lead to compositional structures that may be useful for quantum confinement. Finally, the model performance is discussed, along with recent enhancements that lead to large increases in computational efficiency.

[1] S. Ghosh, D. Kaiser, J. Bonilla, T. Sinno and S. M. Han, Appl. Phys. Lett. 107 072106 (2015).

[2] D. Kaiser, S. Ghosh, S. M. Han and T. Sinno, Mol. Systems Des. & Eng. DOI:10.1039/C6ME00017G (2016).

[3] M. J. Aziz, Applied Physics Letters 70, 2810 (1997).

465803 Substitutional Doping in Nanocrystal Superlattice

Abstract:

Doping, a process in which atomic impurities are intentionally added to modify the electronic properties of a semiconductor, has revolutionized our world, making computers, transistors, detectors, solar cells and other devices possible. Artificial atoms, nanocrystals (NC) with electronic properties dictated by their size and shape, have also emerged as technologically important materials. In this contribution, we introduce the concept of nanocrystal (NC) doping by merging the two above mentioned concepts. We show that, by matching the size of monodisperse gold nanocrystals (Au NCs) with that of semiconducting nanocrystals of cadmium selenide (CdSe) or lead selenide (PbSe) quantum dots (QDs), it is possible to purposely dope the semiconductor superlattices and induce novel properties in these self-assembled materials. We demonstrate that, comparably to the case of atomic doping, the Au NCs take random positions in the superlattice and their concentration can be tuned over a wide range. We show that the electronic and optical properties of the superlattices are affected by the presence of the Au dopants, which increase the conductivity and the photoconductivity of bare CdSe films by several orders of magnitude. We anticipate that this approach can originate a wide variety of NC-doped structures with applications in several fields including electronic materials, solar cells, sensors and catalysis.

456759 Temperature-Dependent Modeling of Formation and Growth of II-VI Semiconductor Nanocrystals

Abstract:

Structural and spectroscopic characteristics of quantum dots (QD) have been experimentally studied for more than a decade owing to their unique physicochemical properties that enable a wide range of applications (e.g. photovoltaics, LED displays, and bio-imaging). Nevertheless, only few studies have attempted to model their formation.[1, 2] Typically, the average size of prepared II-VI QDs can be controlled,[3] but the field lacks the quantitative understanding resulting from comparison between experimental observations (e.g. size distributions and concentrations) and model predictions. Therefore, the aim of this work is to: i) develop a deterministic model describing the formation and growth of II-VI QDs; ii) validate the model against experimental results; and iii) provide insights into the key kinetic parameters and their temperature dependency.

The experimental model system selected is cadmium selenide (CdSe) because of the good control and reproducibility offered by this class of semiconductor nanocrystals. Performing microfluidic experiments *in-situ* absorption spectroscopy, we obtain average time-dependent properties of CdSe QDs (i.e. concentration, average size and distribution width) at temperatures between 160-220°C.[4] Based on population-balance equations, a kinetic deterministic model is developed and the formation of CdSe QDs is described in good agreement with the CdSe experimental data. The present model opens up the possibility to optimize the synthesis of II-VI QDs, while providing temperature dependent kinetic rates for different stages of the QD formation (e.g. nucleation, growth, and dissociation).

Literature:

[1] Abe S, Capek RK, De Geyter B, Hens Z. Reaction Chemistry/Nanocrystal Property Relations in the Hot Injection Synthesis, the Role of the Solute Solubility. Acs Nano. 2013;7:943-9.

[2] Rempel JY, Bawendi MG, Jensen KF. Insights into the Kinetics of Semiconductor Nanocrystal Nucleation and Growth. J Am Chem Soc. 2009;131:4479-89.

[3] van Embden J, Chesman ASR, Jasieniak JJ. The Heat-Up Synthesis of Colloidal Nanocrystals. Chem Mater. 2015;27:2246-85.

[4] Abolhasani M, Coley CW, Xie LS, Chen O, Bawendi MG, Jensen KF. Oscillatory Microprocessor for Growth and in Situ Characterization of Semiconductor Nanocrystals. Chem Mater. 2015;27:6131-8.

469353 Single Enzyme Biomineralization of Size Controlled, Water Soluble Quantum Dots

Abstract:

Single Enzyme Biomineralization of Size Controlled, Water Soluble Quantum Dots

Robert Dunleavy, Leah Spangler, Zhou Yang, Li Lu, Christopher J. Kiely, Bryan W. Berger, Steven McIntosh.

Biological systems have evolved several unique mechanisms to mineralize, primarily structural, inorganic nanomaterials. These biomineralization processes are inherently "green", enabling low-cost and scalable production of nanomaterials under benign conditions in aqueous solutions. However, extending from these processes to achieve the regulated control necessary for reproducible, scalable biosynthesis of functional nanomaterials remains a central challenge. This is especially true of quantum dots (QDs), where precise size and crystal quality control is required to achieve the desired electronic properties. While several studies have described production of QDs from biological systems, the majority of these add a reactive chemical precursor, typically Na_2S, or do not provide the necessary tight control over particle size.

In this work, we describe a single enzyme capable of both the mineralization and templating of cadmium, lead, and zinc sulfide quantum dots within the quantum confined size range. The quantum dots are synthesized in the aqueous phase from normally unreactive metal acetate and L-cysteine precursors and are only formed upon the addition of the enzyme. The mineralization mechanism is proposed to consist of enzymatic turnover of L-cysteine to form reactive H_2S and is coupled to enzymatic templating of nanocrystal growth. We further demonstrate the flexibility of this approach to form core-shell and alloy nanoparticles. The biomineralized materials demonstrate aqueous phase band gap and quantum yield similar to chemically prepared materials.

464838 Microwell Arrays for Screening Interactions Between Root-Associated Microbes

Abstract:

Plant growth promoting rhizobacteria (PGPR) associate with root surfaces in order to provide benefits to the plant host, some of which include improved nutrient uptake, pathogen protection, and plant hormone regulation. An improved understanding of the relationships between PGPR and root hosts has spurred recent interest in developing constructed, artificial communities of PGPR for use as a low-energy, sustainable alternative to chemical fertilizers. However, wide-spread application of constructed communities has been limited due to the fact that the vast majority of microbe-microbe interactions occurring between PGPR in the rhizosphere are unknown or poorly characterized. This limitation stems from the fact that traditional microbiological assays designed to characterize microbial interactions do so using co-culturing methods that are qualitative and low-throughput. Additionally, these assays fail to re-create the physical and chemical complexities of natural ecosystems, such as soil or rhizospere, which are spatially confined microenvironments. To address these types of limitations, we have developed a microfabricated cell-screening platform termed the â€œmicrobe arrayâ€ which uses a novel biofabrication method to randomly assemble thousands of independent bacterial meta-populations into microscale wells with controlled physicochemical properties. It was recently shown that by tuning well dimensions, interacting bacteria can be combined into wells at low dispersity, allowing for the assembly of replicate initial populations; or at high dispersity, allowing for thousands of unique meta-populations to be assembled across a single array of wells. After this, communities can be simultaneously screened for growth or decay to identify combinations of bacteria and environmental parameters that promote or inhibit colonization. In development of this platform, we investigated the effect of spatial confinement on the growth of *Pseudomonas* bacteria. Here, populations were assembled into wells at high dispersity, and growth screening experiments revealed that successful colonization of microbial communities in a spatially confined, diffusion limited environment is highly dependent on inoculum levels. Our current work focuses on screening rhizobacterium populations in this platform to identify pathogenic, competitive or symbiotic interactions that inform the development of constructed communities for biofertilizer and biocontrol applications.

EFFECT OF SIZE AND CHARGE OF METAL IONS ON HYDROGEN PEROXIDE STABILITY IN SILICA HYDROGELS

Ezgi Melis Dogan, Fulya Sudur Zalluhoglu, Nese Orbey, University of Massachusetts Lowell, Lowell, MA, USA

Abstract

Currently, the most efficient method to increase the stability of hydrogen peroxide (H_2O_2) is microencapsulation in silica hydrogels and xerogels via environmentally-friendly sol-gel chemistry using sodium silicate as the starting precursor. The high stability of H_2O_2 in these systems arises from strong hydrogen bonding between silanol groups (Si-OH) and peroxide molecules. Our previous study showed that the amount of Na^+ ions have a strong effect on the structure and morphology of hydrogels. Therefore in the present work, the effect of the size and charge of the ions other than Na^+ on the properties of the hydrogel and stability of peroxide was studied. Monovalent (K^+) and divalent (Mg^{+2}) ions were introduced into the sodium silicate precursor in the form of their respective salts. The pH of the sol was changed by using an ion exchange resin to replace a portion of the ions in the solution with H^+ ions. Gelation time and structure, pH of the sol and stability of H_2O_2 in hydrogels were investigated. Results showed that the stability and gel time can be adjusted to the desired level by changing the ion content and pH of the sol. The increase in total ion concentration in the sol resulted in a decrease in gel time and the best H_2O_2 stability was obtained with the addition of the magnesium salt: seventy percent of the peroxide was retained in hydrogels containing Mg^{2+} ions, as compared to 61 % retention with hydrogels containing Na^+ ions at the end of 70 days. Hydrogels were characterized by using scanning electron microscopy (SEM) and Brunauer-Emmett-Teller (BET) analysis.

Introduction

Hydrogen peroxide (H_2O_2) is an environmentally friendly oxidant which has a limited stability and tend to decompose into oxygen and water according to following reaction.[1]

$$H_2O_{2(aq)} \rightarrow H_2O_{(l)} + \frac{1}{2} O_{2(g)} \tag{1}$$

The concentration of the peroxide, temperature, pH of the aqueous solution, and presence of impurities or stabilizers significantly affects the decomposition rate of H_2O_2. Storage at low temperatures is required for H_2O_2 solutions due to this instability.[2] However, H_2O_2 stability at ambient temperatures is desired by many industrial applications such as sanitation,[3] disinfection,[4] and tooth whitening.[5]

Various methods have been experimented to decrease the decomposition rate and increase the stability of H_2O_2.[6-10]Among these methods, microencapsulation of H_2O_2 in silica hydrogels and xerogels has been the most efficient methods to enhance peroxide stability because of formation strong hydrogen bonds of H_2O_2 molecules with the silica surface.[11]

The synthesis of hydrogels and xerogels is performed using sol-gel chemistry which is a simple and low cost method.[12] The most common precursor of the gels is sodium silicate solution. A colloidal suspension known as sol is first formed by introducing silanol groups into aqueous sodium silicate and removing sodium ions from the solution. For this purpose, a cation exchange resin is used to adjust the sodium content via ion exchange in the solution.[11] This exchange occurs according to following reaction where Rz is the styrene-divinylbenzene copolymer and SO_3H is the sulfonic acid group.

$$Na^+ + RzSO_3^-H^+ \rightarrow H^+ + (RzSO_3)Na \tag{2}$$

A hydrogel is observed in the sol-gel transition point where the viscosity of the solution significantly increase. At this point, Si-O-Si bonds are formed via polymerization of silanol groups in the solution resulting a three-dimensional interconnected network (hydrogel). Xerogels are more contracted gel networks obtained by drying the hydrogels. Silica xerogels containing H_2O_2 has been studied by Żegliński et al.,[6] Bednarz et al.[7] and Wolanow et al..[8] Hydrogel and xerogel properties depend on the drying and storage temperatures, pH of silica sols, and initial precursors. Sudur et al.[9, 10] studied the transition conditions of hydrogel to xerogel and observed the effects of sodium ion on the pH of the sol and H_2O_2 (3.3 to 33.2wt %) stability at room temperature (RT). Significant effects of sodium ion content on the structure, the release rate of entrapped H_2O_2, and the thermal properties of hydrogels and xerogels were found.

The effects of size and charge of ions present in the sol on the structure of hydrogels and stability of entrapped H_2O_2 were questioned because of the observations in our previous work.[9] In the literature, the effect of ion type and concentration on the binding power of sodium silicate was studied for applications in petroleum drilling and construction.[13-15] However, H_2O_2 was not incorporated into the formulations used in these studies. To best of our knowledge, the effect of size and charge of ions on the hydrogel formation or hydrogen peroxide stability has not been studied.

In the present work, H_2O_2 was encapsulated in silica hydrogels using K^+ ions into the sodium silicate precursor in the form of potassium chloride (KCl) at three different molar ratios of K^+: Na^+ (0.14:1, 1:1, and 1.76:1). The effects of K^+ on the gelation time, pH of the sol, stability of H_2O_2 and gel structure were studied. Similar to Na^+, K^+ is monovalent but it has a larger atomic size. On the other hand, silica hydrogels were obtained using Mg^{2+} in the precursor solution to study the effect of charge of the ion on H_2O_2 stability.

Experimental Procedure

Materials

Sodium silicate solution (reagent grade, 10.6% Na_2O, 26.5% SiO_2), hydrogen peroxide (30wt % in H_2O, ACS reagent), Amberlite IR120 (hydrogen form), orthophosphoric acid (85% wt in H_2O); potassium permanganate (ACS reagent >99%) and sulfuric acid (99.999%) were purchased from Sigma Aldrich Chemical Co. Potassium chloride, magnesium chloride and calcium chloride (ACS reagent, >99%) were purchased from J.T. Baker.

Hydrogel synthesis

The procedure of hydrogel synthesis, previously available in the literature[9] was modified. Hydrogels containing Mg^{2+} and K^+ ions along with Na^+ ions were synthesized dissolving magnesium chloride ($MgCl_2$) and potassium chloride (KCl) in deionized water before the addition of sodium silicate. Sodium silicate solution was added dropwise to salt solutions at a volume ratio of 1:5. For hydrogels containing K^+, three different K^+: Na^+ molar ratios of 0.14:1 (low), 1:1 (equimolar), and 1.76:1 (excess) were formulated. In hydrogels containing Mg^{2+}, the molar ratio of Mg^{2+}: Na^+ was adjusted as 0.2:1. Then, all formulations are combined with 30wt % H_2O_2 solution and 7 μL of phosphoric acid (H_3PO_4). The pH of the sols were controlled by adding different amounts (0-3.2 g) of ion exchange resin (Amberlite IR 120, hydrogen form) to replace a portion of the ions in the solution with H^+ ions. The mixture was stirred with an overhead mixer for 4 min. The sol was transferred to a glass Petri dish (D = 100 mm, h = 15 mm) after observing the settlement of the resin. The petri dish was placed in a BlueM gravity oven at 40 °C until observing the gelation.

Characterization

The pH of the sol was changed using different amounts of ion exchange resin at a constant K^+: Na^+ ratio and an initial H_2O_2 concentration (19.9wt %). Gelation time was observed with respect to pH value of the solution for each K^+ formulations. H_2O_2 stability was observed in hydrogels containing K^+ as a function of pH and molar ratio of K^+: Na^+ ions. Stability of H_2O_2 in hydrogels containing Mg^+ was compared based on the presence of ion exchange resin in the starting sol. Hydrogels were stored under ambient conditions at room temperature, and H_2O_2 content was determined using titration method with potassium permanganate ($KMnO_4$) as the titrant. Measurements were made daily for the first 10 days and, then periodically until the concentration of H_2O_2 dropped to 2wt %. Two repeated titrations were performed for each sample and the reproducibility was found to be within 1%.

Quantachrome Autosorb-3B nitrogen sorption porosimeter and JEOL JSM 7401F scanning electron microscope were used to study the morphology and structure of the hydrogels. The surface area and pore size were measured through BET (Brunauer-Emmett-Teller) analysis in the porosimeter. Before these measurements and SEM imaging, aerogels were obtained from hydrogels using a critical-point dryer (Tousimis SAMDRI-795) to preserve the solid network and retain the pore structure. Aerogel samples were coated with gold/palladium target by using a vacuum sputter coater. SEM images of samples were obtained at an acceleration voltage of 10 to 15 kV.

Results and Discussion

Gel time as a function of pH and K^+: Na^+ ratio

Gelation time of low and equimolar of K^+: Na^+ ratio in the sol was studied for different pH values of the sol. Results are plotted in Figure 1. Gelation was most rapid between pH=4 and 6 because the sol passed through the isoelectric point, at which it did not carry a net electrical charge and rapid aggregation of the particles was observed.

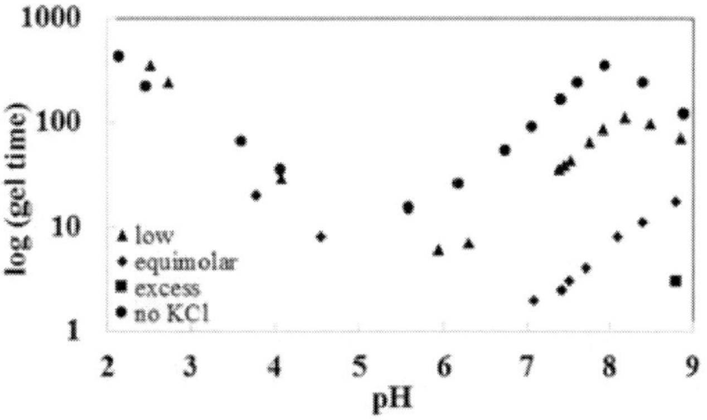

Figure 1: Relationship between gel time and pH for different molar ratios of K^+: Na^+ ions
▲: low (0.14) ♦: equimolar (1) ■: excess (1.76) ●: previous results obtained with unmodified sodium silicate precursor[9]

For all the pH values studied, addition of the ions resulted in faster gelation compared to the previous results obtained in the presence of Na^+ ions only.[6] At the pH range of 2-4, a slow gel formation was resulted because of low concentration of hydroxyl ions. However, longer gel times were obtained after the isoelectric point (pH=6-8) because the gel become negatively charged by high concentration of hydroxyl ions. In the presence of KCl, a shift toward the basic side was observed in Figure 1. This is because the K^+ and Cl^- ions have higher affinity for silicate ions rather than OH^- ions and, therefore more OH^- ions remain in the sol.

Stability of H_2O_2 in hydrogels containing K^+

H_2O_2 content was measured during 70 storage days to study the stability of H_2O_2 as a function of the K^+: Na^+ ratio and the pH of the sol. Stability results of hydrogels containing different K^+: Na^+ ratio at a constant pH of 8.8 are shown in Figure 2. H_2O_2 stability was observed to increase with the addition of K^+ ions as seen more clearly in short term stability (first 10 days during storage).

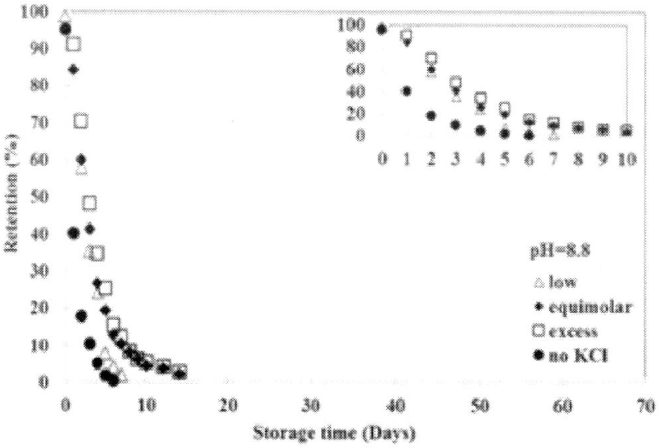

Figure 2: Room temperature stability of H_2O_2 as a function of storage time at constant pH=8.8 (for different molar ratios of K^+: Na^+ Δ: low ♦: equimolar □: excess ●: previous results obtained with unmodified sodium silicate precursor[9])

For example, 18.1wt % peroxide retention was obtained at the end of 2 days of storage at room temperature in our previous work when no K^+ ion was present in the system.[9] In this study, the peroxide retention of hydrogels containing low, equimolar and excess K^+:Na^+ molar ratios was 58.1, 60.2 and 70.6wt %, respectively, during the same period. Moreover, complete decomposition of H_2O_2 in hydrogels containing K^+ took longer (7 to 14 days) than that in previously studied hydrogels[9] (5 days).

Figure 3 a & b show the effects of pH on peroxide stability for low and equimolar amounts of K^+: Na^+, respectively. For each ratio, when the pH of the sol decreased, higher retention of H_2O_2 was observed. At lower pH values of the sol, higher amounts of H^+ ions increase the entrapment efficiency and retention by creating more available sites for peroxide in silica network.[16]

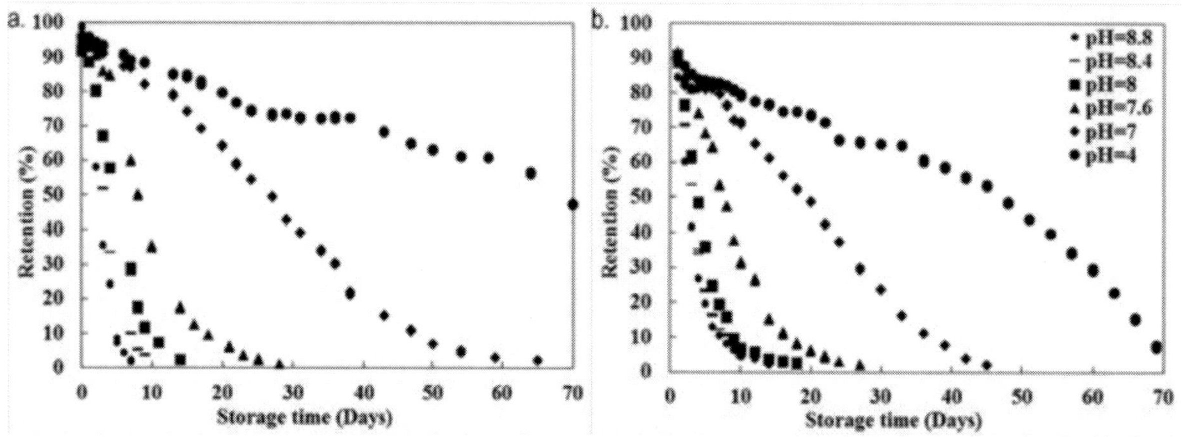

Figure 3. The comparison of H_2O_2 stability in hydrogels containing low (a) and equimolar (b) ratios of K^+: Na^+ at different pH values

Stability of H_2O_2 in hydrogels containing Mg^{2+}

H_2O_2 stability of hydrogels containing Mg^{2+} was studied to observe the effect of ion charge added into the starting silicate solution. Magnesium was selected to be partially exchanged with the sodium in the silicate at 1:0.2 molar ratio of Mg^{2+}: Na^+ because magnesium ion is divalent and smaller in size than Na^+. Magnesium ion could not be completely replaced with Na^+ because magnesium silicates are not water soluble unlike potassium silicates. The following reaction shows the precipitation reaction of magnesium silicates.

$$2(\equiv SiO^-Na^+) + MgCl_2 \rightarrow 2 \equiv SiO^-Mg^{2+} + 2NaCl \qquad (3)$$

The pH of the sol containing Mg^{2+} was decreased using ion exchange resin (Amberlite IR120) which replaces ions with H^+. Figure 4 shows H_2O_2 retention in hydrogels synthesized in the presence (pH=4.6) and absence (pH=8.8) of ion exchange. Previous stability results of hydrogels containing only Na^+ are also included in Figure 4 for comparison. The presence of Mg^{+2} showed improved peroxide stability regardless of the presence of ion exchange. We believe that the reason of this result is more sites available for hydrogen bonding between silanol groups and peroxide molecules due to the smaller size and higher charge of magnesium ions than Na^+.

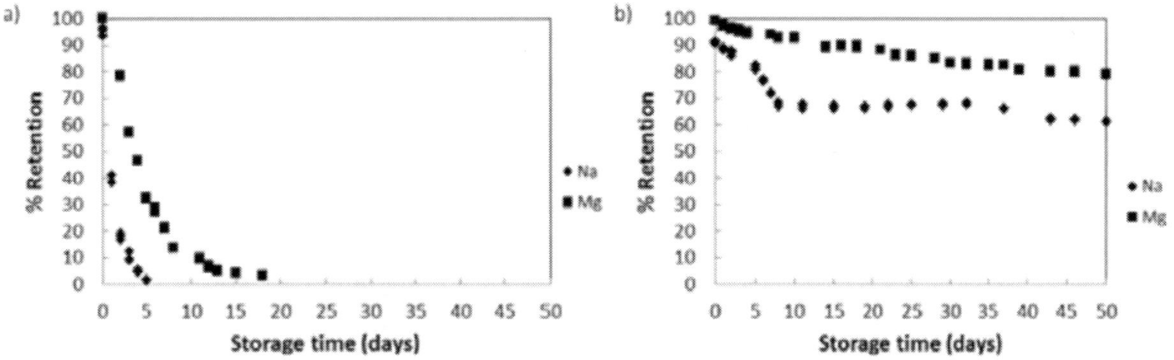

Figure 4. Room temperature stability of H_2O_2 as a function of storage time (a) without ion exchange (pH=8.8) and (b) with ion exchange (pH=4.6)

Morphology and structure of hydrogel

BET and SEM analyses showed that the presence of potassium ion resulted less compact gel structure at constant pH. This is because K^+ has larger atomic size than Na^+ and results larger pores in the hydrogel structure. Figure 5 shows SEM images of samples having different molar ratios of K^+: Na^+ at pH=8.8.

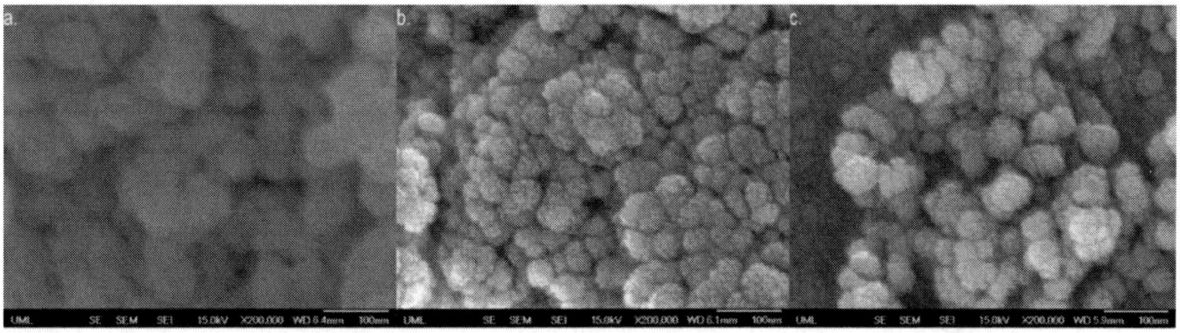

Figure 5. SEM images (at X200,000) of hydrogels for different molar ratios of K^+: Na^+ : a) low, b) equimolar, c) excess (at constant pH=8.8)

Nitrogen sorption studies were carried out to obtain quantitative results of the effect of potassium ion on pore size and surface area of hydrogels. Results are shown in Figure 6. When K^+: Na^+ molar ratio was increased, larger pore sizes were obtained because of the larger atomic size of K^+ (Figure 6a). For each case, small pore sizes were observed as pH decreased because hydrogen ions have capability to bring silica particles together by charge screening effect. In Figure 6b, a decrease in surface area was resulted in the presence of K^+. However, surface area of hydrogels increased at lower pH values as expected due to the decrease in pore size.

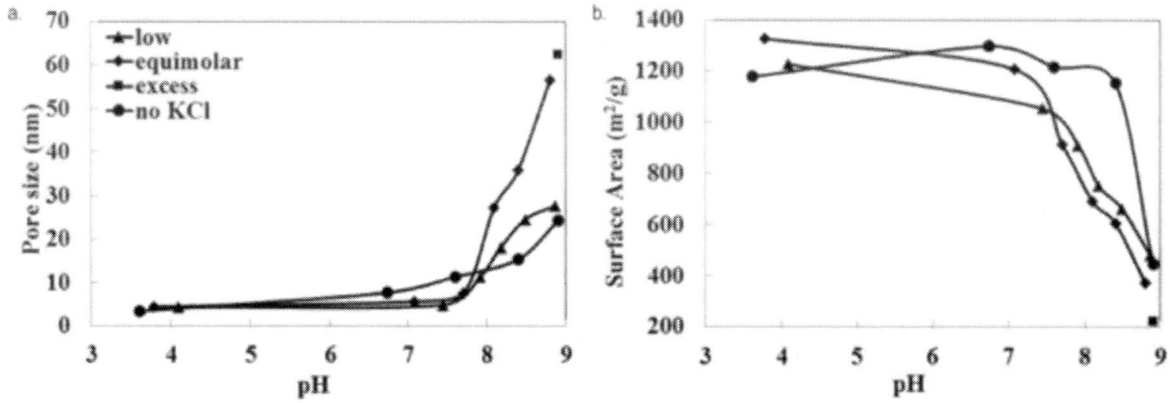

Figure 6. Structural properties of silica hydrogels with changing pH and K^+: Na^+ molar ratio, ▲: low ♦: equimolar ■: excess ●: previous results obtained with unmodified sodium silicate precursor[9]

Conclusions

A significant effect of the size and charge of ions on H_2O_2 stability and the hydrogel structure was observed. Gelation time decreased as K^+: Na^+ ratio were increased in the starting precursor. The stability of encapsulated H_2O_2 was improved with the addition of Mg^{2+} and K^+ ions and decreasing pH of the sol. With the increase in K^+: Na^+ ratio, average pore diameter was found to increase and surface area decreased at constant pH (8.8), therefore the gel became less compact. The higher retention and better stability of H_2O_2 at room temperature was resulted by controlling the ion content (size and charge) and pH. The properties of hydrogels can be tailored to control the gel time and H_2O_2 stability. A detailed study might be required for better understanding of the effect of size and charge of ions on peroxide stability and the hydrogel structure. For example, calcium ions can be included in hydrogel synthesis because the sizes of Ca^{2+} and Na^+ ions are similar whereas it has higher charge like Mg^{2+}.

References

1. Schumb WC, Satterfield CN, Wentworth RL. Hydrogen peroxide. ACS Monograph Series. New York: Reinhold Publishing Corporation, 1955.
2. Sigma Aldrich Product Information Sheet: Hydrogen Peroxide Solution. http://www.sigmaaldrich.com/
3. Colodette JL, Rothenberg S, Dence CW. Factors Effecting Hydrogen Peroxide Stability in the Brightening of Mechanical and Chemimechanical Pulps. J. Pulp. Pap. Sci. 1989; 15: 3–10.
4. Sulieman M, Addy M, MacDonald E, Rees JS. The Effect of Hydrogen Peroxide Concentration on the Outcome of Tooth Whitening: An In Vitro Study. J. Dent. 2004; 32: 295–299.
5. Kim JC, Lee KU, Shin WC, Lee HY, Kim JD, Kim YC, Tae G, Lee KY, Lee SJ, Kim JD. Monoolein Cubic Phases Containing Hydrogen Peroxide. Colloids and Surfaces B: Biointerfaces. 2004; 36: 161–166.

6. Żegliński J, Cabaj A, Strankowski, M, Czerniak J, Haponiuk JT. Silica Xerogel-Hydrogen Peroxide Composites: Their Morphology, Stability, and Antimicrobial Activity. Colloids and Surfaces B: Biointerfaces. 2007; 54: 165–172.
7. Bednarz S, Rys B, Bogdal D. Application of Hydrogen Peroxide Encapsulated in Silica Xerogels to Oxidation Reactions. Molecules. 2012; 17: 8068-8078.
8. Wolanov Y, Shurki A, Prikhodchenko PV, Tripolskaya TA, Novotortsev VM, Pedahzur R, Lev O. Aqueous Stability of Alumina and Silica Perhydrate Hydrogels: Experiments and Computations. Dalton Trans. 2014; 43: 16614.
9. Sudur F, Pleskowicz B, Orbey N. Hydrogen Peroxide Stability in Silica Hydrogels. Ind. Eng. Chem. Res. 2015; 54 (6):1930–1940.
10. Sudur F, Orbey N. Properties of Hydrogen Peroxide Encapsulated in Silica Hydrogels and Xerogels. Ind. Eng. Chem. Res. 2015; 54(44): 11251-11257.
11. Bergna HE, Roberts WO. Colloidal Silica: Fundamentals and Applications. Boca Raton, FL: CRC Press, 2005.
12. Baccile N, Babonneau F, Thomas B, Coradin T. Introducing Ecodesign in Silica Sol-Gel Materials. J. Mater. Chem. 2009; 19: 8537–8559.
13. Brady AP, Brown AG, Huff H. J. The polymerization of aqueous potassium silicate solutions. Colloid Sci.1953; 8(2):252-276.
14. Shroff AV, Shah DL. Grouting Technology in Tunneling and Dam Construction. Rotterdam, Brookfield: Oxford & IBH Publishing Company, 1992.
15. Chun-xi Z. Recent Advances in Water glass Sand Technologies. Shanghai, China: School of Material Science and Engineering, Shanghai Jiaotong University; 2007.
16. 27. Iler RK. The Chemistry of Silica: Solubility, Polymerization, Colloid and Surface Properties and Biochemistry. New York: Wiley, 1979.

472182 UV-Assisted Synthesis of Carbon Nanotube-TiO_2 Nanocomposites for Enhanced Photocatalytic Air Purification

Abstract:

Efficient photocatalysts consisting of carbon nanotubes (CNTs) decorated with TiO2 nanoparticles were synthesized by a simple two-step process that is free of aggressive acid treatments that often introduce defects to CNTs. The novel synthesis process involved UV-assisted oxidation of CNTs using H_2O_2, followed by the deposition of TiO2 via a wet chemical method using TiCl4 as a titanium source. The resulting CNT/TiO2 nanocomposites have been characterized by X-ray diffraction (XRD), transmission electron microscopy (TEM), X-ray photoelectron spectroscopy (XPS), and Raman spectroscopy. XRD results revealed that the crystalline structure of TiO2 is anatase and the TEM results showed the undamaged tubular morphology of CNTs decorated with TiO2 nanoparticles. The XPS data revealed the intimate chemical interactions between the CNTs and TiO2 in the nanocomposites. The expected synergistic effect and efficient separation of photogenerated electron-hole pairs in the synthesized CNT/TiO2 nanocomposites were verified in gas-phase photocatalytic degradation of VOCs.

470129 NOVEL Magnetic Nanocomposite Materials for RAPID Removal of Polychlorinated Biphenyls from Contaminated Water Sources

Abstract:

Magnetic nanocomposite systems able to capture polychlorinated biphenyls (PCBs) in aqueous solution were developed, providing a cost-effective water remediation technique. Two distinct methods were employed to develop these polyphenolic nanocomposite materials, where the polyphenolic moieties were incorporated to create high affinity binding sites for organic pollutants. The first method utilized a surface initiated polymerization of polyphenolic-based crosslinkers and co-monomers on the surface of iron oxide magnetic nanoparticles to create a core-shell nanocomposite. The second method utilized a bulk polymerization method to create macroscale films composed of iron oxide nanoparticles incorporated into a polyphenolic-based polymer matrix, which were then were processed into microparticles. Both methods produce nanocomposite materials that can bind chlorinated organics, can rapidly separate bound organics from contaminated water sources using magnetic decantation, and can use thermal destabilization of the polymer matrix for contaminant release and material regeneration. The polyphenol functionalities used to bind organic pollutants were quercetin multiacrylate (QMA) and curcumin multiacrylate (CMA), which are acrylated forms of the nutrient polyphenols with expected affinity for chlorinated organics. Particles were characterized using transmission electron microscopy (TEM), dynamic light scattering (DLS), Fourier transform infrared spectroscopy (FTIR), thermal gravimetric analysis (TGA) and UV-visible spectroscopy stability analysis. Pollutant binding studies were performed using PCB 126 as a model PCB and chlorinated organic pollutant to determine binding affinity and capacity, and this was quantified using gas chromatography coupled to electron capture detector (GC-ECD). It was demonstrated that the materials effectively bound PCBs in aqueous media.

472163 Influence of Silica-Based NanoparticlesÂ Embedded in Sand Bed Filtration for Cleaning-up Industrial Wastewater

Abstract:

Sand filtration technology is widely used to remove the suspended solids and colloids from the industrial wastewater effluent. However, such standalone technology cannot remove the soluble organic compounds from the industrial wastewater. In the current application, we are targeting to improve not only the removal of suspended solids but also increase the removal efficiency of total organic carbons (TOC) by enhancing the sand surface area via integrating it with functionalized silica-based nanoparticles. These kinds of nanoparticles are environmentally friendly and naturally occurring materials at the macro-level.

In this study, different types of functionalized silica-based nanoparticles were successfully prepared, characterized, and tested for TOC adsorptive removal within batch and continuous adsorption processes. Briefly, the Brunauerâ€"Emmettâ€"Teller (BET) surface area, X-ray diffraction (XRD), Fourier transform infrared (FTIR) spectroscopy and high resolution transmition electron microscopy (HRTEM) were used to confirm the surface area, crystalline domain size, functionality, and structural topology, respectively. The characterization results showed that the prepared nanoparticles are spherical in shape with a crystalline diameter of 10 nm and BET surface area around 180 m^2/g. Adsorption experiments were conducted in a batch mode to investigate the adsorption affinity and capacity of nanoparticles toward TOC removal from a locally supplied industrial wastewater sample. The results indicated that the TOC removal efficiency was improved significantly by 50% using the functionalized sand with nanoparticles at low mass fraction (<10 wt%). Afterwards, the filter bed column, continuous mode, was conducted using the functionalized and the un-functionalized sands. Interestingly enough, the functionalized sand with in-house prepared silica-based nanoparticles showed prolonger breakthrough time (>50 times) in comparison with the un-functionalized sand. These promising results are believed to help the current conventional wastewater treatment methods not only on a lab scale

but also on an industrial level.

464632 Inactivation of *E. coli.* Using a Novel TiO_2 Nanotube Electrode

Abstract:

Electrochemical technology has gained increasing interests for water/wastewater treatment in recent years. In particular, it may find applications in decentralized treatment due to its simplicity, amenability to automation, compact instrumentation, and ease of operation. The performance of electrochemical processes highly depends on the material of electrodes. However, the high cost of electrodes, formation of toxic byproducts, and short electrode lifetime have impeded its wide application. Therefore, development of cost-effective anode materials are needed to support future adoption of electrochemical technologies for environmental applications. In our study, a novel TiO_2 nanotube anode has been fabricated in the lab via simple electrochemical anodization using metallic Ti as the substrate followed by electrochemical hydrogenation/cathodic polarization. The one-dimensional nanostructure is shown in FESEM images. The TiO_2 nanotube anodes with stainless steel as the cathodes were tested for inactivation of a model microbial contaminant *E. Coli.* using a filter-press type cell. Effects of current density applied (0.625 ~ 2.5 mA/cm^2), electrolyte concentration (0.3 mM~3 mM NaCl), and different electrolyte (NaCl, Na_2SO_4, Na_2HPO_4, $NaNO_3$, and $NaHCO_3$) on *E. Coli.* inactivation were investigated. Electrochemically generated oxidizing species were measured as free chlorine. Further, production of reactive oxygen species (•OH, O_2•-, SO_4^{2-}•) were examined using probe compounds. In addition, electro-disinfection experiments were conducted in real water matrices, i.e. samples of surface water, treated surface water before disinfection, and secondary effluent before disinfection collected from local water/ wastewater treatment plants. Formation of inorganic (e.g. ClO_3^-) and organic (THMs) disinfection byproducts were also monitored. Our results have demonstrated that the novel TiO_2 nanotube anode can effectively achieve 5~6 log reduction of E. Coli. with low byproducts formation. The low energy consumption also indicates that such disinfection system may be

operated using solar energy stored in a DC battery for decentralized treatment or as a point-of-use device.

451725 Metal Organic Framework Derived Nanoporous Carbon Â As a Novel Adsorbent for Water Treatment

Abstract:

Metal organic framework derived nanoporous carbon

as a novel adsorbent for water treatment

Zahra Abbasi[a], Ezzatollah Shamsaei[a], Soo Kwan Leong[a], Bradley Ladewig[a,b], Xiwang Zhang[a], Huanting Wang[a*]

[a]Department of Chemical Engineering, Monash University, Clayton, Victoria 3800, Australia

[b]Department of Chemical Engineering, Imperial College London, Exhibition Road, London SW7 2AZ, United Kingdom

Phone: +61 3 9905 3449, Email: Huanting.wang@monash.edu

Nanoporous carbon materials have attracted a great deal of attention because of their potential application in many areas such as adsorption and catalysis. These materials have desirable properties such as high surface area, good chemical and thermal stabilities as well as fast kinetics, which make them suitable for adsorption applications. Porous metal organic frameworks (MOFs) consist of crystalline structures that provide high internal surface areas and tunable porosities which could be considered as starting materials for developing novel carbon based nano composite

adsorbents.

Different types of adsorbents have been studied for the removal of organic pollutants from water and wastewater such as activated carbon, fly ash waste materials and different kind of MOFs. However, their adsorption capacity is low in comparison with nanoporous carbons. Therefore, developing novel carbon based materials with simple synthesis methods for adsorption of pollutants from water is of great importance.

Zeolitic immidazolate frameworks (ZIFs) are a class of MOFs that exhibit remarkable features like high thermal and chemical stability and demonstrate similar pore topologies to zeolites. Among various ZIFs reported, ZIF-8 with the formula of $Zn(2\text{-methylimidazolate})_2$ and with a sodalite-related zeolitic structure has been used for many applications such as adsorption and catalysis. The surface and bulk properties of ZIF-8 can be modified to make it a more competent adsorbent for water treatment purposes. Carbonization of ZIF-8 could be the easiest and most practical modification approach to amend this metal organic frame work's characteristics.

The adsorption capabilities and characteristics of nanoporous carbon particles derived from ZIF-8 were investigated in this study. Direct carbonization of ZIF-8 was used to synthesize carbon nano particles. SEM and TEM images revealed that the ZIF-8 derived carbon carbonized at several temperatures retained the original structure and morphology of ZIF-8. These carbon nano-particles exhibited significantly greater adsorption capacities for the removal of dye from water compared to the ZIF-8. The high adsorption capacity is attributed to the change in surface charge and pore size. The surface functionality and hydrophobicity of the materials were investigated by Raman Spectroscopy, N_2 adsorption-desorption, FTIR, TGA and contact angle measurements. The change in the surface charge of the carbon particles was confirmed by zeta potential of the samples. The dye uptake followed pseudo-second order kinetics and the equilibrium data was best fit to the Langmuir model for the carbon nano-particles. Intra-particle diffusion studies showed that the adsorption process was governed by both the diffusion step and external mass transfer. The ZIF-8 derived nano porous carbons were found to be highly efficient adsorbents for water and waste water treatment.

471195 Study of TL Response of Silver Nanoparticles in Borate Glasses Containing Dy^{3+} and Eu^{3+} Ions for UV and Gamma Dosimetry

Abstract:

Abstract

Dy^{3+} and Eu^{3+} doped the borate glasses samples containing Ag nanoparticles were prepared using a conventional melt-quenching method. The synthesized materials were characterized by scanning electron microscopy (SEM) and their morphology and sized distributions were determined. The photoluminescence (PL) spectra as well as UV-Vis absorbance were also analyzed and show the presence of the rare-earth in the matrix. The thermoluminescence (TL) response of Dy^{3+} doped lithium borate glasses irradiated with ultraviolet (UV) radiation was studied in detail before and after doping with silver.

471313 Effects in Morphology and Thermoluminescent Characteristics of Lif Crystals Synthesized By Using Nonionic and Cationic Surfactants

Abstract:

LiF crystals doped with Ce^{3+} and Mg were prepared using co-precipitation, using as solvent a mixture of ethanol and water in presence of Nonionic (PVP and Pluronic f-127) and cationic (CTAB) surfactants. The LiF crystals were characterized by scanning electron microscopy (SEM) obtaining average sizes of 100 nm, a single cubic crystalline phase was determined by XRD, the PL emission spectra shows two bands characteristics of Ce^{3+} and UV-Vis absorbance shows the presence of the rare-earth in the LiF crystals. The TL response to both X-ray and gamma from a Cs-137 source irradiation was studied.

470030 Thermoluminescent Dosimetric Analysis of Ag and Cu Doped LiB3O5

Abstract:

Methods

Production

Synthesis of Ag and Cu doped LiB_3O_5 crystals was carried out by precipitation, using water as a solvent. Impure crystals were produced at a synthesis temperature of 70 °C. A 5g of Ag and Cu doped LiB_3O_5 precipitate was obtained upon evaporation of water in the solution. The precipitates were calcined at a temperature of 750 °C for one hour. The calcined materials were grounded finely and subjected to severalcharacterizations.

Characterization

SEM analyses were performed for the doped and undoped samples of LiB_3O_5. The XRD properties of the samples were obtained. The UV-Vis absorbance spectra of the samples were obtained at room temperature. Spectra were recorded between 200 and 1200 nm. The photoluminescence (PL) measurements were carried out. The glow curves of the samples has been obtained using TLD reader.

Results

The XRD patterns obtained for the samples showedthat production of LiB_3O_5 with orthorhombic structure was successful since allof the main peaks of LiB_3O_5 coincide with the JCPDS data. It was seen that the dopant added did

not interfere withthe crystal structure. As it can easily be detected addition of the dopants didnot have any significant effect on the crystal structure of LiB_3O_5 produced.

SEM was carried out to check the morphologies of the samples.According to the SEM images, the samples produced by the co-precipitation method shows the agglomerated surface.

The UV-Vis absorption spectra for the samples showed that the maximum absorbance was observed at 250nm. The result showed that a clear monotonic and increasing dependence of the intensity of the absorbance as a function of dopants.

The dosimetric properties of LiB_3O_5 doped with Ag and Cu were determined by using HarshawTLD Reader Model 3500. The heating rate was chosen as 10°C/s was usedfrom 50 to 350°C. The examined dosimeters were exposed to X-rays at room temperature. The sample amount was 0.5g. The samples have shown the TL response. The pattern of the glow curves changed with the change of dopants.

Conclusions

The TL response of LiB_3O_5 has been compared with the commercial TLD100. The observation concludes that the doped LiB_3O_5 has the TL behavior. The glow curves of the LiB_3O_5 shows different shape with the change of dopants.

469662 Effect of Thermal Treatment on the Characteristics of PES/PVA Nanocomposite Membranes Modified with TiO2 Nanoparticles :a Comparitive Study Between 1-Step and 2-Step Thermal Treatment

Abstract:

PES/PVA thin film composites were modified with TiO_2 nanoparticles via dip coating method. Thermal treatment of total modified membranes plays a semi-"catalytic role" in coating process. Hence, 4-temprature levels were selected and investigated, noticing that there is an important restriction due to the stability limitation of polymers under thermal treatment process. This phenomenon also leads to the concept of 2-step treatment. 2-step thermal treatment is a more moderate process which leads to a more stable structure.

Polyethersulfone (PES) ultrafiltration membrane was prepared via phase inversion induced by immersion precipitation. Poly(vinyl-alcohol) (PVA) (1.5 wt%) dissolved in distillated water was coated on the surface of PES barrier layer with dip coating technique. The PVA-coated membranes were then immersed in the cross-linking solution containing 5 wt.% of GA and 0.5 wt.% of H_2SO_4 for 2min at room temperature in order to reduce the membrane swelling by chemical cross-linking. H_2SO_4 was used as catalyst of reaction. TiO_2 nanoparticles were dispersed in distillated water by sonication for 1hr and stirred vigorously for 30 min. the PVA coated membranes were immersed in 0.2 wt% TiO_2 solution for 15 min. the last step is thermal treatment which in this study, four level of temperature (85,95.105.115) were investigated. All modified membranes were thermal treated for 3 min.

The performance of modified membranes was tested with filtration of yeast industrial wastewater(COD=11500 mg/L). Permeate flux and COD of four modified membranes have tested. According to the results, the PES/PVA/TiO_2 membrane that was thermal treated in 105 had the best output.

469788 Synthesis, Structural and Morphological Characterization of Cu and Ag-Doped $Li_2B_4O_7$

Abstract:

Method:

In this paper the lithium tetra borate (LTB) was produced by water/solution assisted synthesis method. After the synthesis, metals (doping with Copper (Cu) and Silver (Ag)) were doped into lithium tetra borate ($Li_2B_4O_7$) to enhance its thermo luminescent properties. The heating temperature was 750°C for 2 hr as well as 150°C for 1hr the sample produced by water solution method with 0.1% Cu and 0.1% Ag. The pellets were irradiating by the X-ray.

Result:

The identification and characteristics of undoped and doped $Li_2B_4O_7$ were determined by X-ray diffraction (XRD), Scanning electron microscope (SEM), Photoluminescence (PL), Ultra violet visible spectroscopy (UV Vis). The chemical compositition and its morphologies of the obtained LTB and LTB: Cu, Ag was confirmed by XRD and SEM results. The most intense peak of the XRD pattern of the lithium tetraborate sample is verified by the reference data of the lithium tetra borate with tetragonal structure. The glow curves were obtained by thermo luminescence (TL) reader. The glow curves of the pellets exposed to different doses exhibited a clear response to X-ray irradiation, which is a significant evidence that this material is a candidate for radiation dosimetry. Finally this energy response was also compared with TLD100. The comparison of the results showed that the obtained lithium tetra borate have good thermo luminescence (TL) dose response.

Enhancement of Heat Transfer Coefficient by Using Fe_2O_3 –Water Nanofluids

N.I. Zouli[1], Muthanna Al-Dahhan[1]
[1]Missouri University of Science and Technology, Rolla, MO, 65409
Email: niz8vb@mst.edu

Introduction

Research efforts have been made to increase the heat transfer rate, energy, and efficiency. Some of these efforts include the application of fins, compact heat exchangers, microchannel, and many others. Recently, nanofluids have been introduced to enhance the heat transfer rate of the heat transfer fluids by adding a certain concentration of nanoparticles into the heat transfer media. Most of the work that has been published in the literature focuses on measuring the thermal conductivity of the nanofluid or measuring the overall heat transfer coefficient. Among the efforts for enhancement of heat transfer the application of additives to liquids is noticeable (Bergles, 1973; Ahuja, 1975). Choi (1995) was the first who studied the enhancement of the nanoparticles on the heat transfer coefficient and showed that the nanoparticles could considerably affect the thermal conductivity of the nanofluid to increase. Since Choi is work (Choi,1995), various studies have been conducted on nanofluids and these studies shown that nanofluids increase heat transfer properties and stability while reducing costs, clogging, and energy required to pump the fluid. Heris et al. (2006) found that the heat transfer coefficient increased as the volume concentration of the nanofluids increased when the fluid was composed of water and metal oxides.

In the present study, effects of Fe_2O_3/water on the nanofluid's heat transfer and thermal conductivity will be discussed. The experimental investigations are done to study heat transfer performance of the nanofluid at the entrance region under laminar flow condition varying the particle size and volume concentration of nanoparticles.

Experimental setup and methodology

Schematic of the experimental setup used in the current study is shown in Figure1. It consists of a flow loop made of copper material, a heating unit, a measuring and control unit. A straight copper tube with 190.5 mm length, 25.4 ± 0.02 mm inner diameter, and 31 ± 0.05 mm outer diameter was used as the test section. The heat transfer probe comprises of a heat flux sensor (0.15 in x0.08 in) with a fast response time of about 0.02 sec and a thermal resistance of 0.003 °F/BTU/ft2hr. The sensor was placed in a slit mounted inside the test section wall flashed to the wall without disturbing the flow. The probe measures the heat flux and the surface temperature. In addition, two thermocouples (type T) were mounted inside the test section to measure the bulk temperature. The sensor was placed on the middle distance between the two thermocouples.

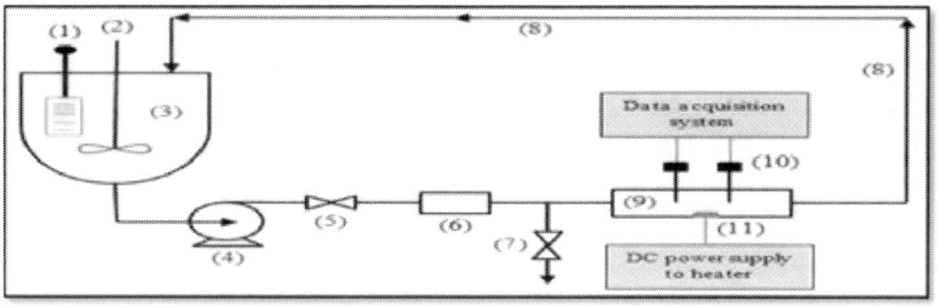

Figure 1. Schematic of experimental setup components: (1) Heater, (2) Agitator, (3) Reservoir tank, (4) Pump, (5) Needle valve, (6) Flow meter, (7) Drainage valve, (8) Copper tube, (9) Test section of copper tube, (10) Thermocouples, (11) Heat flux foil sensor.

Heat flux, temperature of the sensor surface, and the bulk media temperature are directly measured and recorded using Data acquisition system. Further they are used to estimate local instantaneous and averaged heat transfer coefficient using eqn (1) and (2):

$$h_i = \frac{q_i}{T_{si} - T_{bi}} \; ; \; \Delta T_i = T_{si} - T_{bi} \qquad (1)$$

$$h = \frac{1}{n} \sum_{i=1}^{n} \frac{q_i}{T_{si} - T_{bi}} = \frac{1}{n} \sum_{i=1}^{n} h_i \qquad (2)$$

where h_i: local instantaneous heat transfer coefficient ($kW/m^2.K$)
q_i: instantaneous heat flux measured by the sensor (kW/m^2)
T_{si}: instantaneous surface temperature of the probe (K)
T_{bi}: instantaneous bulk temperature of the media (K)

Results and discussion

The thermal conductivity of Fe_2O_3 water based nanofluids was measured at different concentrations. From Figure 2 an enhancement of 22.35% in the thermal conductivity value has been achieved with 0.09% volume concentration of Fe_2O_3 nanoparticles in water with an average diameter of 20 nm. As can be seen from Figure 2, the thermal conductivity enhancement increases linearly with increased particle volume concentration.

The concentrations for each test run were increased from 0.02-0.09% by volume and run at a constant flow rate of 6 L/min. The experimental results were obtained with 20 nm particles. And it was found that the average increase in the heat transfer coefficient was between 5.95-18.34% with the highest increase being 73.7% at 0.09% volume concentration at 5 mins into the process. It was observed that as the concentration of the nanofluid was increased so did the heat transfer coefficient. The enhancement in heat transfer coefficient was calculated by following;

$$[(h_{nf} - h_{bf}) \,/\, h_{bf}] * 100\% \qquad (3)$$

h_{nf}= heat transfer rate of nanofluid
h_{bf}= heat transfer rate of base fluid

A typical result for 20 nm particles can be seen in Figure 3. Overall the results show that the local heat transfer coefficient increases with increasing the size and the volume fractions of the Fe_2O_3 nanoparticles compared to water. The effect of Reynolds number on Nusselt number is shown in Figure 4. It is seen that the measured Nusselt numbers for nanofluids are higher than those of base fluid, and they increase remarkably with Reynolds number. The observed enhancement of the Nusselt number could be due to the suppression of the boundary layer, the viscosity of nanofluids as well as the dispersion of the nanoparticles.

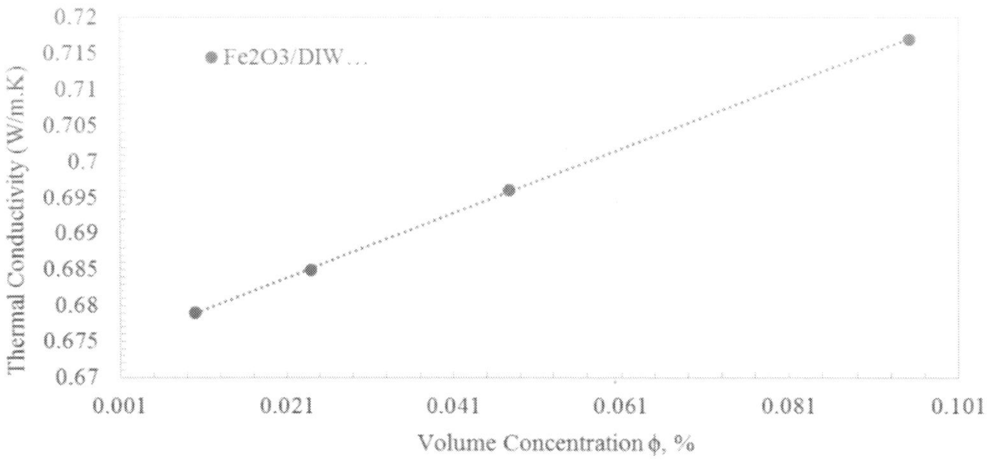

Figure 2. Thermal conductivity of Fe_2O_3 water based nanofluids at different volume fractions (0 and 0.09 %).

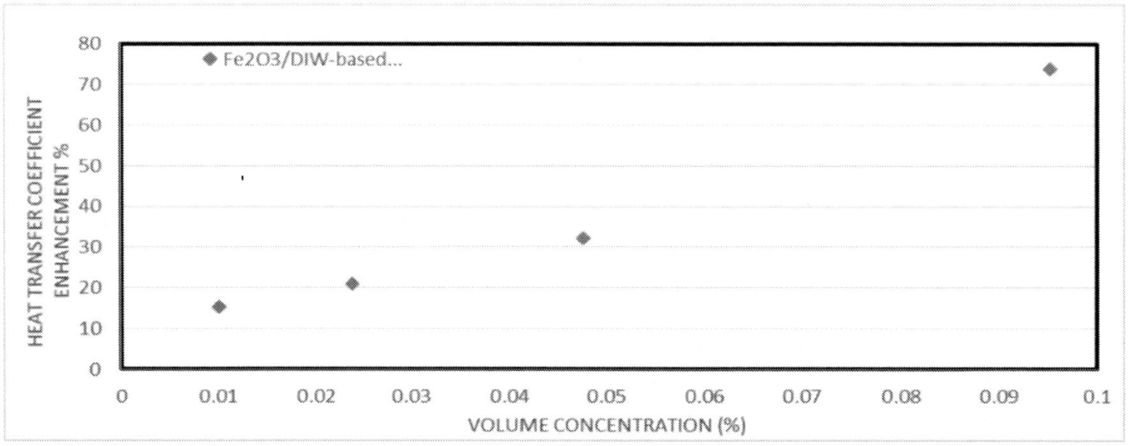

Figure 3. Variation of heat transfer coefficient with different volume concentration of Fe_2O_3 water based nanofluids.

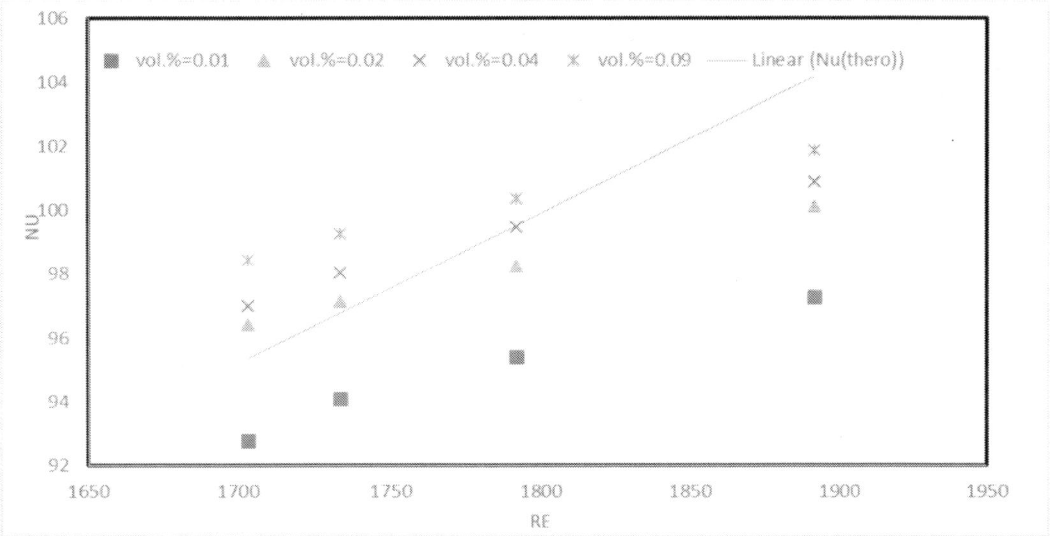

Figure 4. Effect of Reynolds number and Fe_2O_3 nanoparticle concentration on Nusselt number

Conclusions

The paper is mainly concerned with the heat transfer of nanofluids made of water and Fe_2O_3 nanoparticles in the laminar flow regime. It was found that use of aqueous Fe_2O_3 nanoparticles can significantly enhance heat transfer coefficient in laminar flow regime, and the enhancement increases with Reynolds number, as well as particle concentration under the conditions of this work. The enhancement of the heat transfer coefficient could not be only credited to the enhancement of the effective thermal conductivity. A reason for the enhancement, Particle passage which lead to reduce the thermal boundary layer thickness and a non-uniform distribution of thermal conductivity.

Acknowledgments

I would like to acknowledge the financial support provided by the government of Saudi Arabia. I would also like to thank Prof. Al-Dahhan for his supervision and support which made this work possible.

References

1. Brooke, A., Kendrick, D., Meeraus, A. A. (1992). GAMS- A User's Guide (Release 2.25). *The Scientific Press*. San Francisco, CA.

2. Gooding, W. B., Pekny, J. F., McCroskey, P. S. (1994). Enumerative Approaches to Parallel Flowshop Scheduling via Problem Transformation. *Comput. Chem. Eng.*, *18*, 909.

3. OSL. (1991). *OSL, Guide and Reference (release 2)*, IBM: Kingston, NY.

4. Zhang, X., Sargent, R. W. H. (1994). The Optimal Operation of Mixed Production Facilities - A General Formulation and some Approaches for the Solution. *In Proceedings of the 5th International Symposium on Process Systems Engineering*. Kyongju, Korea, 171.

5. A.E. Bergles, Recent development in convective heat transfer augmentation, Appl. Mech. Rev. 26 (1973) 675–682.

6. A.S. Ahuja, Augmentation of heat transport in laminar flow of polystyrene suspension: I. Experimental and results, J. Appl. Phys. 46 (8) (1975) 3408–3416.

7. Choi, S.U.S. Enhancing thermal conductivity of fluids with nanoparticles. in American Society of Mechanical Engineers, Fluids Engineering Division (Publication) FED. 1995.

8. Heris, S.Z., S.G. Etemad, and M.N. Esfahany, *Experimental investigation of oxide nanofluids laminar flow convective heat transfer*. International Communications in Heat and Mass Transfer, 2006. **33**(4): p. 529-535.

447375 Eco-Friendly Dyeing of Electrospun Cellulose Nanofibers with Reactive Dye Using Ultrasonic Energy

Abstract:

Nanofibers have created enormous opportunities and challenges for the textile and apparel industry, moreover the innovative methods for dyeing of nanofibers have recently gained more attention. Electrospun cellulose nanofibers are increasing significant interest in textile industry in view of its incomparable properties. However, the dyeing of nanofibers faces problem of lower color yield due to the higher surface area of nanofibers. In this study electrospun cellulose nanofiber dyed with using cold pad batch dyeing, a well-known and ecofriendly method of dyeing for cotton base fibers. The dyeing of cellulose nanofibers was carried out with CI Reactive Violet 5 by conventional and ultrasonic (US) method. Scanning electron microscopy and Fourier transform infrared spectroscopy were used to investigate morphology and chemical structure of electrospun cellulose nanofibers before and after dyeing and the color strength (K/S value) and color fastness of dyed nanofibers also assessed.

The results indicates that the use of ultrasonic energy not only shortens the batching time but the color strength (K/S),color fixation and color fastness also considerably increased ,the K/S value of dyed samples enhanced around 5-8% compared with conventional method. Although, washing fastness of sample increased up to 25%, but, color fastness to light was the same as conventional method.

The results proved that, the morphology of dyed cellulose nanofibers with both dyeing method were smooth with structural uniformity and remained unaffected after being dyed.

452910 Combined Quartz Crystal Microbalance with Dissipation and Generalized Ellipsometry to Characterize the Deposition of Titanium Dioxide Nanoparticles on Model Rough Surfaces

Abstract:

Titanium dioxide nanoparticles (TiO_2NPs) are the most extensively manufactured engineered nanomaterials, and understanding of their interactions with soil and sandy aquifer materials is necessary for effective natural resource management [1]. Quartz crystal microbalance with dissipation monitoring (QCM-D) has been widely used to study TiO_2NPs at water-metal oxide interfaces as model systems for soils and aquifers. However, natural sands and sediments exhibit various degrees of surface roughness, and the effects of surface roughness are not accounted for in current QCM-D studies of nanoparticle adsorption, which use the Voigt-Kelvin model [2].

A new mechanical model that considers both viscoelastic and surface roughness effects [3] is described to analyze QCM-D measurements. Obtained parameters include adsorbate areal mass density, adsorbate viscoelastic properties, the vertical roughness length, and the permeability length. Silicon slanted columnar thin films (SCTFs) were deposited onto QCM-D sensors by electron-beam glancing angle deposition and conformably coated with an alumina film via atomic layer deposition to act as model rough surfaces. The adsorption of polyacrylate-stabilized TiO_2NPs onto flat surfaces and the SCTFs was measured by QCM-D and generalized ellipsometry (GE), an optical reference technique.

We report that the surface-roughness-modified mechanical model excludes water that is retained by the SCTF surface roughness and instead considers the surface mass density of TiO_2NPs and water coupled to the TiO_2NPs [4]. GE is also sensitive to the attachment of the TiO_2NPs. When the results of both techniques are combined, new information, the hydration of the polyacrylate-stabilized TiO_2NPs, can be evaluated.

[1] B. Kim et al. J. Environ. Monit., **14**, 1128 (2012).

[2] M.V. Voinova et al. Phys. Scr., **59**, 391 (1999).

[3] L. Daikhin et al. Anal. Chem., **74**, 554 (2002).

[4] N. Kananizadeh et al. J. Hazard. Mater., In press. dx.doi.org/10.1016/j.jhazmat.2016.03.048

AUTHOR INDEX

AUTHOR INDEX

AUTHOR INDEX

AUTHOR INDEX